ARMAMENT AND TECHNOLOGY

ARTILLERY AND MISSILES

Illustrations: Octavio Díez Cámara, Daimler-Benz Aerospace, Empresa Nacional Santa Bárbara, GIAT Industries, Lochkeed Martin, Matra BAe Dynamics, MLRS International Corporation, Oerlikon Aerospace, Oerlikon Contraves Defence, Raytheon Electronic Systems, Rheinmental W & M, Saab Dynamics, SITECSA, Thomson-CSF Airsys, United Defense and Wegmann & Co.

Production: Lema Publications, S.L.
Editorial Director: Josep M. Parramón Homs
Text: Octavio Díez
Editor: Eva Mª Durán
Coordination: Eduardo Hernández
Translation: Mike Roberts
Original Title: Artillería y misiles

ISBN 84-95323-29-X

Photocomposition and photomecanics: Novasis, S.A.L.
Barcelona (Spain)
Printed in Spain

ARMAMENT AND TECHNOLOGY

ARTILLERY AND MISSILES

LEMA
Publications

The British started using the 105-millimeter OTO Melara howitzer in 1959, a weapon that was used to great effect in such diverse places as Aden and Borneo. Despite being extremely sturdy, its range was rather limited for many military requirements, and this became more and more evident as the years went by. A new model was developed to replace it, one that would combine advanced design, the lightness that was needed to move it around easily and a barrel that would be long enough to reach targets at distances of up to twelve miles. The result was the L118 Light Gun that has been used in several places of conflict, such as the Falkland Islands, the Persian Gulf and Bosnia.

Designed by and for artillerymen

The engineers at the Royal Armament Research and Development Establishment at Fort Halstead began work in 1966, basing the project on the requirements of the British Army in 1965 for an artillery weapon that was both light and stable, had a long range and that could be towed across difficult terrain at high speeds.

PREPARATION OF FUSES

The different types of 105-millimeter projectiles must be suited to both the fuse and the propulsion charges depending on the distance of the target and the type of damage that must be inflicted upon it.

DISCREET USE

In this picture, the barrel is retracting after firing a shot, and we can see that one of the outstanding qualities of the Light Gun is the discretion of both its muzzle and the amount of dust that the firing provokes.

The prototypes are tested

The prototypes that the Development Establishment produced were subjected to a wide variety of tests; in 1973 the 105/37-mm gun was incorporated, given the codename L118 and known popularly as the "Light Gun". Construction began at the Nottingham Royal

Ordinance factory, and the first gun was ready in October 1974. Almost two hundred more soon followed and they gradually took the places of the similar models that the Royal Artillery, the Royal Marines and the Air Force had used up until then, artillery groups with three batteries of six pieces.

PREPARING FOR FIRE

The users of a 105/30 millimeter gun, hidden under a mimetic net to make it harder for the enemy to locate it, insert the projectile and the propulsion charge into the breech, and in a matter of seconds the gun is ready to fire again.

In 1981, Australia chose this gun, and bought 111 of them, naming it the "Hamel Gun" and they were produced under license by Australian Defense Industries Limited. In early 1986, the United States Army decided to adopt this British weapon, referring to it as the M119. The weapon was essentially the same L119 but with a shorter gun so as to allow for the use of the large 105-mm munitions that were available at the time. The first 147s were produced by the Royal Ordinance and the rest (an expected 583 to be made under license) by the Rock Island Arsenal at Illinois and the Watervliest Arsenal in New York.

A GOOD METHOD OF SELF-DEFENSE

With the barrel at an angle of 0°, the 105/30-millimeter gun becomes an excellent weapon for defense against advancing forces; it can defeat a modern tank or armored vehicle situated over half a mile away.

Among the other users of this model –of which more than a thousand have been made to date as well as several more under license in other countries– are Abu Dhabi, who bought 59, six of which were passed on to the Dutch Marines in 1995; Bahrain, which uses 8; Botswana, with 6; both the Brazilian Army and Marines, which use 36 and 4 respectively; Dubai, who bought 59 along with Abu Dhabi; Spain, with a total of 56 that are used in the three brigades of its Fuerza de Acción Rápida (FAC – Rapid Action Force), and refer to them as the L118A1 and the L119A1 because they use both the version with the longer gun and the shorter one; Ireland, who have 12; Kenya, who have 40; Malawi, with 12; Morocco, with a total of 30; New Zealand, who received 24 from Australia; Oman has 39 and Portugal, the latest country to show interest, ordered twenty-one L119 in early 1997.

Used to liberate the Falkland Islands

The enormous military force that Argentina had supposedly sent to the Falkland Islands with the aim of claiming the islands as their own, caused the British Royal Artillery to transport five batteries into the heart of the area that the "Task Force" had established itself in ready to liberate the islands. There were thirty 105 mm Light Guns that were about to make their first appearance in a real-life military operation. After the landing at San Carlos on the 21st of May, 1982, three batteries of six L118A1 guns belonging to the 29th Command Regiment and one battery belonging to the fourth Campaign Regiment were deployed.

These weapons (which proved to be extremely accurate and have a remarkably long range) were first used by the 8th Command Battery, who used three of theirs from a position at Camilla Creek House in support of the 2nd Battalion of the Parachute Regiment during actions in Darwin and Goose Green. Many of the guns were transported

by Westland Sea King or CH-47 Chinook helicopters to assist with the movement and relocation of artillery. All of them played important roles in the conflict; the 6,000 rounds that were fired in just 12 hours during the final assaults on Tumbleweed, Wireless Ridge and Mount William provide testimony of the fact.

Outstanding characteristics

Designed to satisfy the requirements of the last years of the 20th century, but hoped to still be in service throughout the first few decades of the 21st, the Light Gun fulfills the needs of most modern artillery. It can be used without limitations in such different environments as desert sands, the Amazon rainforest or the frozen north of Norway, and it can either be towed along by light trucks or even in the sling of medium type helicopters.

Constituent elements

This gun –produced in both its standard form and as the modernized A1 type– is outstanding for its elevating mass, made up of a hydropnuematic recoil shock absorber, an electric activator for magnetic type firing and the barrel itself. The latter is an L19 piece with a length of 3.21 meters with a rifled interior with 28 grooves, which has been fitted with a two-stage muzzle at the front. If the user wants, a 30 caliber, 2.779-meter barrel can be mounted for firing older M-1 type munitions, which are percussion activated and have a maximum range of 11,500 meters.

The former is moved on a steel base, at the front of which is the suspension and two 9.0 by 16 wheels, fitted with their own hydraulic brake which allows the gun to be towed along at extremely high speeds. The two supports are fitted at the rear, designed to allow the gun to be towed and also to stabilize it during firing (which can not only be

HIGHLY MOBILE

The lightness of the carriage, less than two tons, and its wheels provide the mobility that allows it to towed by light trucks and vehicles, which also transport the soldiers and munitions so it can be used immediately.

FAST AND EFFECTIVE DEPLOYMENT

The artillerymen of the "Almogávar" Parachute Brigade of the Spanish Ground Force use three L118A1 six-gun batteries that can be transformed into the L119A1 model. They can deploy the guns from transport aircraft by using parachutes.

configured for both direct and indirect pointing, but also for direct emergency aiming, in which case the gun is operated by a single man, for whom there is a special seat). The base plate is transported on the barrel's two thick twin-support carriages when the gun is in its towing position (in which case the barrel faces backwards). This plate assists the turning of the weapon when it has to be aimed at targets in different directions.

Seven charges can be used with the L19 barrel of the L119 configuration, reaching overlapping ranges of distance, allow for the targeting of objects at distances between 1.5 and 10.75 miles. The L27 barrel and the L127 gun were adopted to satisfy the requirements of Switzerland, which received six guns up until 1981, though this was a number that fell short of the total that had originally been ordered. The Spanish Land Force, for example, has received guns with two barrels, which, assisted by a small crane, can be changed in less than an hour. Consequently, they can be used to fire older munitions for training procedures and be adapted for more modern ones in the case of their being used in real-life combat.

Options for the future

The United States has perfected the L119A1 LASIP (Light Artillery System Improvement Plan), which includes a redesigned base plate, a more robust firing control system and a new element for direct firing, a recuperator designed for use in extremely cold conditions of up to 50° below zero and that does not require as much maintenance, and new muzzles that are more appropriate for when the gun is towed by light HMMWV type vehicles. M913 HERA (High Explosive Rocket Assisted) munitions are being developed along with M915 DPICM (Dual Purpose Improved Conventional Munitions) to maximize its characteristics, firing capacity and range (which is currently about 19,500 meters when using the assisted type).

The British have also updated the weapon in several ways, and since early 1999 have arranged for, after extensive international trials, the purchase of 137 APS systems (Automatic Pointing System) at a cost of 6.1 million US dollars to be installed on their Royal Ordnance L118 guns (which are usually towed by Steyr-Daimler-Punch Pinzgauer Turbo D 4x4 vehicles).

These models, which shall also receive the name LINARP (Laser Inertial Automatic Pointing), will be made by Marconi Electronic Systems and production should be complete by the year 2002. They are basically made up of an inertial system and a GPS (Global Positioning System), an LDCU display screen, an odometer and an energy system that is used for controlling it.

ARTILLERY LEGIONNAIRES

The legionnaires of the Spanish King Alfonso XIII Legion Artillery Group continually train for the deployment and use of the eighteen guns they own, and take full advantage of its wide range of uses.

MODERN AND EFFICIENT

Although it has been in service for twenty-five years, this British gun is still used as the basis for all later adaptations, which says a lot for its capacity and features.

ELECTRIC OR MANUAL FIRING

The barrel is fitted at the rear with the breech where the projectiles are positioned before being fired. This is electric type firing with an L19 barrel that can use advanced munitions or manual ones with the L20 for more traditional munitions.

RECOIL ELEMENTS

Two buffers that combine springs and hydropneumatic elements are what cause the barrel to return to its original position after firing, meaning that the recoil of the weapon is only 0.33 meters at an angle of 70° and 1.07 meters when firing at 0°.

TWIN SUPPORT CARRIAGE

The twin support at the rear, made of a thick and hollow steel tubes in a fixed semi-circular position, not only allows for towing but also for seating. Three or four men are needed to move the gun.

WHEELS

British Light Guns have large 9.0 by 16 wheels with built-in braking systems, which means that they can be towed across all types of surfaces at considerable speeds.

DIRECT AND INDIRECT AIMING

On the left of the Light Gun is a complex himing goniometer that makes for more precise indirect firing and includes an element that makes the gun usable for self-defense under difficult conditions by means of tense firing.

RECOIL ELEMENT

A one or two stage muzzle can be incorporated (there are openings on either side) at the front of the barrel to reduce recoil in the firing movement, minimize flare at the mouth and divert gasses that could make the gun easier for the enemy to spot.

SHORTER BARREL

The L119A1 are the result of fitting the L118 with a 30 caliber L20 barrel which can fire older munitions, which include all the American M-1 types.

LARGE BASE PLATE

Positioned above the twin-support carriage for transporting, or under the two wheels in the firing position, the plate has an unusual shape, which stops it sinking into soft surfaces after firing. It also makes it much easier to make 360° turns.

TECHNICAL CHARACTERISTICS LIGHT GUN L118A1

COST IN DOLLARS:	1,070,000 including spare parts, short barrel and munitions
CALIBER:	105 mm
DIMENSIONS:	
Length in firing position	6.629 m
Length in transport position	4.876 m
Height	1.7778 m
Width	1.371 m with the barrel in transport position and 2.63 in that of firing
Height off ground	0.50 m
WEIGHTS:	
Prepared for combat	1,860 kg
Fuel	1,066 kg of the lifting mass and 794 kg of the support and other elements
FEATURES:	
Range with minimum load	2,500 m
Range with maximum load	17,200 m
Angle of elevation/depression	+70°/-5.5°
Rate of fire	9 rounds per minute and 3 in the case of sustained fire
CREW:	5 men

The need to supply the five Italian mountain units and one air transport brigade, with the artillery that they required (for tasks such as the movement of the former across the mountains in the north of Italy and the deployment of the latter in more varied environments). Led the National Army to request the specialized Italian industry to develop a new piece of light caliber artillery.

Its features, capacity to be disassembled into 11 elements that the mountain and air transport troops can transport easily, compact size and remarkable mobility have meant that this model has been produced in great numbers. More than 2,500 units have already been made and have been exported to some thirty different countries.

Introduced very quickly

The first requirements for a piece of 105 millimeter caliber artillery were manifested by the Italian Ground Forces in the early fifties, and the first versions were ready for testing midway through that decade. In 1957, having proved the gun's excellent potential, the OTO Melara de la Spezia company (known today as OTO Breda) started manufacturing it, and it was not long before orders were arriving for all corners of the globe.

HIGHLY MOBILE

Its weight of 1,290 kilograms means it can be transported by medium type helicopters. Heavier ones, such as the CH-47 Chinook in this photo, can transport two of them in the sling and the crew and munitions can be placed in the helicopter's fuselage.

VERY LITTLE RECOIL

The configuration of the mount, the power of the muzzle and the characteristics of the munitions it uses mean that the recoil is minimal. This is the moment that this photo shows, just after the shell has left the barrel.

Many armies decide to adopt it

The need for a light artillery element that could be easily deployed to wherever it might be needed, led to several military organizations in a wide variety of different countries to express an interest in this model. For example, more than a hundred were produced for the Parachute, Air Transport, Mountain and Legionnaire Brigades of the Spanish

Ejército de Tierra, where it is still used as support for light infantry units. At the same time, the BRIMAR (Brigada de Infanteria de Marina) of the Spanish Navy have two batteries of six of these guns which are used as basic support weapons on beachhead operations.

Other users of this Italian weapon, which was later named M-56 in reference to the year that production was given the go-ahead, are Saudi Arabia, Argentina (who deployed some of them in their occupation of the Falkland Islands and were used to defend the islands from the British offensive), Bangladesh, Brazil, who ordered twenty of them in 1995, Canada, where they were called the C5, Chile, Djibouti, Ecuador, Germany, Great Britain, who deployed them in Borneo and South Yemen, Greece, India, Iraq, Kuwait, whose units were captured by the Iraqis early on in the Gulf War, Malaysia, Morocco, Nigeria, Peru, Portugal, Somalia, Sudan, Thailand, Venezuela, Yugoslavia, Zambia and Zimbabwe.

The Chinese copies

In the nineties, Alenia Difesa, a division of OTO Breda, sent two models of their 105-mm howitzer to China, but has not received any further orders. In mid 1997 it was found that a company called NORINCO (China North Industries Corporation) was promoting a light weapon that could also be used as an anti-tank weapon, which was almost identical in design and caliber to the Italian design.

It can use standard United States type projectiles and can be used to launch M1 HE (High Explosives) at a distance of 10,222 meters. This weapon follows the trend of other Chinese weapons that are identical or very similar to western designs, which are then sold worldwide at extremely competitive prices, although a few changes are made to the design to get around accusations of plagiarism. However, it seems that the PLA (People's Liberation Army) is not interested in this howitzer and there is no confirmed evidence of any of them having been exported.

WIDE VARIETY OF MUNITIONS

The design was optimized for using the complete range of United State 105 millimeter munitions that are produced in many different countries; amongst these is an HE projectile that can be used against any kind of target.

Highly mobile in any circumstance

When the M56 was designed, special attention was given to the need for the possibility of disassembling the gun so that it could be transported to less accessible places. Therefore it was configured as eleven basic elements that could be dismounted easily and taken, for example, on the backs of donkeys up steep mountain passes or put in the holds of transport helicopters in those situations where it would not be appropriate to carry the gun in the sling.

Simple and efficient configuration

The 56 model incorporates a very short

COMPACT SIZE

The M56 105 millimeter weapon is noted for its compact size and light weight of just 1,290 kilograms, which makes it easier to hide under mimetic nets and also assists with its transportation, which can be done by many different types of vehicle.

DEPLOYED IN BOSNIA

The Spanish mountain troops, part of the 1st Aragon Brigade, were deployed in Bosnia along with their M56 howitzers in case they were needed to support their humanitarian actions; these weapons tend to employ a protective shield.

14 caliber gun that measures 1.478 meters and has a 1.074 meter rifled bore in its interior (with 26 grooves), capable of making the round turn for a maximum distance of 10,000 meters with the seventh projection charge, made up of 1,323 grams of propulsion explosive. The barrel has a useful life of 7,500 rounds using the maximum charge. At the front it has a multiple defector muzzle that assures that most of the gasses produced by firing are dispersed to the sides, thus making the weapon harder to locate and reducing recoil; at the same time, it has a hydraulic buffer and a spring-operated recuperator that reduces the recoil of the barrel to a range that varies from 240 millimeters to a maximum of 190. The shutter, which has a breech volume of 2.057 cubic decimeters and withstands pressure of up to 1,930 kg/cm², plugs vertically; the gun carriage is twin-masted and is combined differential deformation type and its wheels have two 7.00 x 16 CEAT safety type tires with pressure of 1.4 kg/cm². These allow the weapon to be towed along at high speeds and are related to an articulated suspension with torsion bars; the width of the wheels varies from 1.32 meters, the most normal size, to a minimum of 1.14, depending on the surface on which the weapon is to be moved. The axis of the trunnions can be fitted in a high position in which the firing sector is from +65° in elevation to –5° in depression, or in a lower anti-tank position in which the firing muzzle has an elevation of +25° and the same depression; the height is reduced dramatically from 1.93 meters to 1.55, making it harder to locate and making movement easier.

USED WORLDWIDE

More than 2,500 units of the M56 have been bought by thirty countries, who have admired the robustness of this Italian weapon, its compact size and its capability for performing highly complex firing actions; it can be towed by medium size vehicles.

Different characteristics

The possibility of disassembling it into different parts means that the 118.2 kg shield is usually removed for normal activities that do not need this kind of protection against enemy fire. Other basic elements include the barrel, the muzzle, which weighs 31.3 kg, a sledge that is used for moving across snowy or muddy surfaces, the gun carriage, the aiming elements and the two wheels, which weigh 64.6 kg each. Also worthy of mention are the two supports that can be made up of two or three sections depending on tactical needs; to be towed by land vehicles it can be configured in a position of 3.65 meters and for being towed by animals this is raised to 5.3 m; to assist with aiming it includes a 105/14 type sight, a Righi double graduation level, a direct pointing eyeglass of 1.8 magnification and with an amplitude of 150 thousandths for using HEAT grenades and a double graduation, 1.5 magnification Salmoranghi panoramic goniometer; its vertical fire sector goes from 0 to 1,155 thousandths on the upper axis and from 89 to 533 on the basic axis; the horizontal sector is from 640° on the first and from 995° on the second. Therefore, it can be towed by a light 0.75 ton four wheel drive vehicle or carried in the sling of a UH-1H medium type transport helicopter, positive factors in making this an extremely versatile weapon for troops to use on deployment maneuvers.

CREW

Seven men in normal artillery units and nine in mountain units form the crew that transports, positions and fires the light italian howitzer, whose users admire its robust nature and easy maintenance.

EASY TO HIDE

A metal support and a medium sized mimetic net hide the gun from enemy observation –its compact forms and small size assist with this need.

Highly varied munitions

The M56 was designed for firing the same munitions as those used by the United States' M101 and M105 105 mm towed guns, munitions that are produced all over the world and are 100% compatible with this Italian weapon, unlike many other more modern weapons. Of the former there are High Explosive (HE) power munitions that are made up of a 21.06 kg projectile, which have an initial speed of 472 meters a second and that, once it has exploded, can affect an area with a radius of 10 x 15 m; one HEAT perforator for anti-tank fire that weighs 16.7 kg and pierces 102 mm of armor in its M67 configuration; several smoke type munitions, both white smoke for smothering or others in brighter colors for indicating specific positions; one illuminating type called M314 for lighting selected areas during the night, that of reduced charge instruction known as CSST, that for anti-tank instruction called ETRL, the H M60 for launching chemical products and the BE M84, which is used for sending different types of propaganda. For firing, there are a total of nine charges that vary from 258 to 1,323 grams of projection gunpowder and with initial speeds of from 180 to 472 m/s.

AIMING ELEMENTS

On the left hand side of the weapon are the different aiming elements, including a 1.5 magnification panoramic goniometer, a fixed sight, a double graduation level and a direct pointing eyepiece, not particularly sophisticated elements, but extremely reliable.

MUZZLE

A robust, 33.1 kilogram muzzle can be fitted to the front of the gun, which includes five chambers on each side to disperse the gasses that accompany the projectile towards the mouth.

STRONG BREECH

The M56's shutter includes a 2.507 cubic decimeter breech into which the projectile and the propulsion charge in a tin pod are inserted. Its vertical plugging is easy to activate and notably solid.

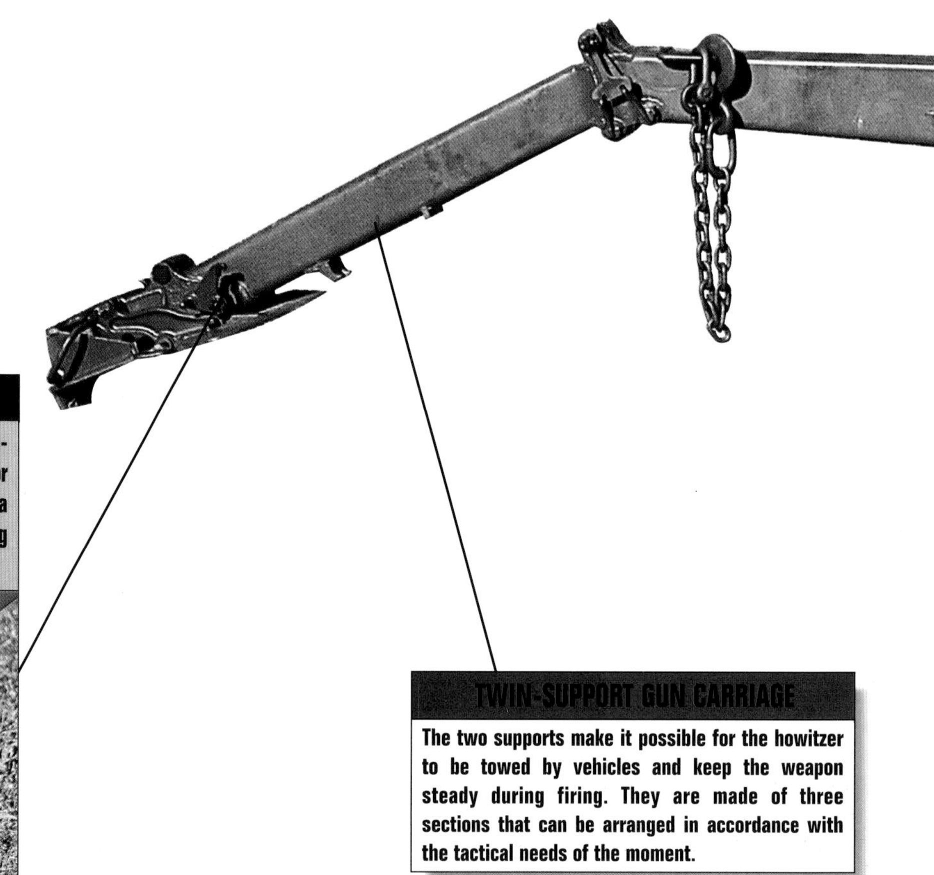

DISMOUNTABLE SUPPORTS

The two supports of this Italian weapon can be dismounted into three pieces each to make the piece shorter or assist with transport. At the end of the last section is a grid that helps it grip to the ground and stay still during firing.

TWIN-SUPPORT GUN CARRIAGE

The two supports make it possible for the howitzer to be towed by vehicles and keep the weapon steady during firing. They are made of three sections that can be arranged in accordance with the tactical needs of the moment.

TECHNICAL CHARACTERISTICS TOWED HOWITZER MOD. 56 105/14

COST IN DOLLARS:	600,000
CALIBER:	105 mm
Length in firing position	4.8 m
Length in transport position	3.65 m
Height	1.93 m
Width	1.5 m
WEIGHTS:	
Total in combat configuration with two part support	1,290 kg

FEATURES:	
Range with minimum load	2,900 m
Range with maximum load	10,575 m
Angle of elevation/depression	+65º/-5º
Firing rate	8 rounds per minute and, sustained for an hour, 3 shots
CREW:	
7 men for campaign units and 9 for mountain units	

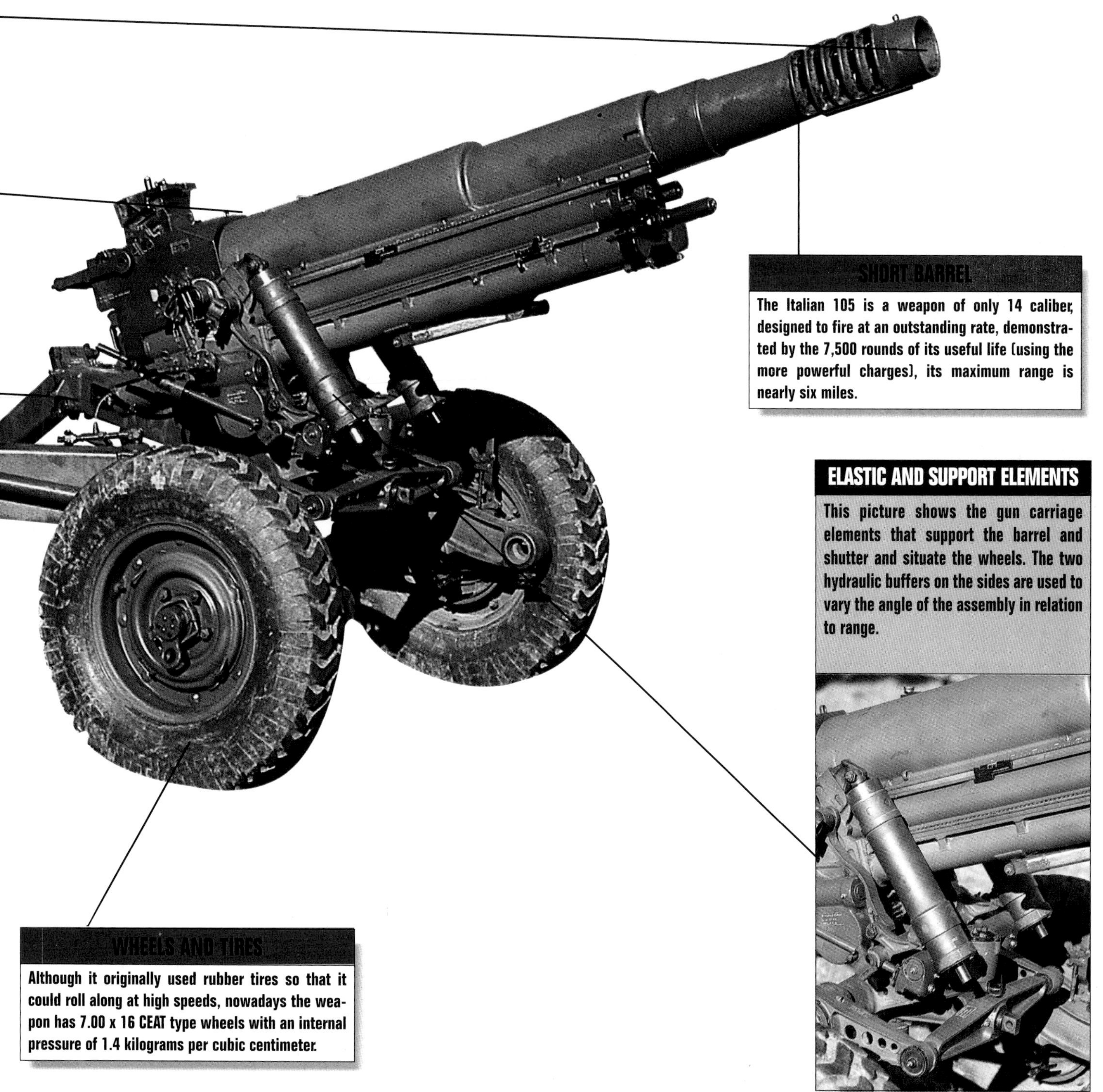

SHORT BARREL

The Italian 105 is a weapon of only 14 caliber, designed to fire at an outstanding rate, demonstrated by the 7,500 rounds of its useful life (using the more powerful charges), its maximum range is nearly six miles.

ELASTIC AND SUPPORT ELEMENTS

This picture shows the gun carriage elements that support the barrel and shutter and situate the wheels. The two hydraulic buffers on the sides are used to vary the angle of the assembly in relation to range.

WHEELS AND TIRES

Although it originally used rubber tires so that it could roll along at high speeds, nowadays the weapon has 7.00 x 16 CEAT type wheels with an internal pressure of 1.4 kilograms per cubic centimeter.

There was a clear need to supply the different towable artillery units of the Spanish, Ground Forces of the Artillería de Campaña (ACA) with new weapons, and several models of towed howitzers of 155 and 203 millimeters were put into development. The prototypes have since appeared and have been subjected to careful testing.

The current state of many of the artillery batteries (which still use the very old Naval Reinosa R/58 105/26 mm howitzers, the 105/14 mm Oto-Melara M-56 and the United States 155/23 mm M-114) suggests the need for several improvement and modernization plans. These plans include the adoption of the British Light Gun and the exhaustive evaluation of other Spanish proposals for higher caliber and greater capacity weapons, with the aim of reducing artillery stores and configuring weapons with greater fire power, both with respect to firing speed and range.

CHINESE MADE BARRELS

The barrels of the FGT-203/45 gun were provided by NORINCO. They are 45 calibers long, with which it could reach distances of 30 miles when using "base bleed" type projectiles.

Privately funded developments

A company called Desarrollos de Sistemas y Tecnologias S.A. (SITECSA – Systems and Technology Developments), belonging to the Explosivos Río Tinto (ERT) group, set up, in 1986, a project that aimed to commercialize different weapon systems. These included new artillery weapons that needed to stand up to the requirements of the Spanish Ground Forces.

GOOD TACTICAL MOBILITY

Although the FGT-203/45 gun weighs more than 16 tons, its is highly mobile, and a 6x6 truck could tow it at a maximum speed of 50 mph when used on roads, and 18 mph when used on unmade surfaces.

Imported technology and its own capacity

A good starting point would be the different studies of ballistics and the artillery weapons produced between 1960 and 1969 by a Canadian engineer called Gerry V.Bull –a key feature of the United States HARP (High Altitude Research Program) project, the SRC group, based in Brussels, designed the 155 mm GC 45 howitzer.

The technology and plans that this project office produced, along with the acquisition of an Austrian weapon known as Noricum N-45, permitted work on different projects that aimed to create three different models of medium caliber weapons with 39, 45 and 52 caliber barrels that could or could not incorporate Auxiliary Power Units (APU). The prototypes were made at an industrial workshop at Granollers, near Barcelona, and included several towed howitzers, including the 155/45 mm ST-01.2, and the ST-01.2b, which differed from the first in that it is able to move by itself thanks to the incorporation of a 130 horse-

power APU diesel engine.

The staff of the Artillery Academy of the Spanish Ministry of Defense put these gun carriages to several tests in places that ranged from the National Maneuvers and Firing Range at San Gregorio in Zaragoza to the Costilla Artillery Range in Cadiz. The APU movements proved to be extremely successful, allowing for speeds of up to 20 mph; its maximum range was 40,000 m with ERFB-BB munitions, and its accuracy varied just 0.2% in range and 0.4 thousandths in direction.

Of particular note is the 7.046 m barrel with a rifled bore and a constant rate of 20 caliber; the screw breech; the elastic organ made up of a hydraulic brake and an oleopneumatic recuperator; the tubular pneumatic fastener; the gun carriage made of solded steel; the wheels that are configured in two independent sections with two wheels on each and with box-shaped steel supports; it used to weigh 9 tons and in the firing position it was 11.4 meters long.

The company was also involved in the possible updating of the M-114, a low budget scheme that aimed to increase the availability of the weapon and improve some of its features; a prototype was completed with a 155 mm and 45 caliber barrel and with several minor changes made to its elastic organs.

MOVEMENT ON ALL SURFACES

The 155/45-millimeter SITECSA ST-01.2b howitzer was very well made and in the tests that the Spanish Ground Forces subjected it to, it proved capable of moving on all surfaces.

TECHNICAL FEATURES SB 155/52 APU SBT-1 HOWITZER

COST IN DOLLARS:	1,200,000
CALIBER:	155/52 mm
VOLUME OF THE CHAMBER:	23 liters
DIMENSIONS:	
Length in firing position:	12.7 m
Length in transport position:	10.8 m
Length of barrel:	8.12 m
Height:	2.3 m
Width:	2.8 m

WEIGHTS:	
Total in combat order:	12,900 kg
FEATURES:	
Range with maximum load	40,000 m
Angle of elevation/depression	+72°/-3°
Firing rate	3 rounds per minute
MOBILITY:	
APU configured of a 106 HP four-speed diesel engine that moves it at 11 mph	
CREW:	4 men

The Chinese connection

An agreement signed at the end of the eighties between SITECSA and the Chinese NORINCO company gave the go-ahead to a project called Bull that worked on a towed weapon that would be named FGT-203/45 (Field Gun Towed 203 mm). As a result of this collaboration, the latter provided two 203 mm barrels, two muzzles and two breechs, elements that the former integrated, along with its own mount, in a factory based at Segovia. The process took from 1988 to July 1990 and required an investment of thirteen million dollars.

The firing tests took place at the end of that year and the weapon was unveiled to the Army at the Fuencarral Artillery Academy in Madrid. It was later sent to China, just as had been accorded by the mutually signed contract. Its most significant characteristics include the fact that it has an extremely light mount that can launch projectiles over distances of up to 30 miles. For this, it incorporates a type of technology called "managing steel", that allows for the manufacture of a highly resistant steel alloy with elements such as carbon and cobalt. The use of this material reduced the overall weight by some 3,000 kilograms; its weight was 16,396 kg.

It was 14.831 meters long in the firing position, 2.8 meters high, and could be towed along roads at a speed of 50 miles per hour. It needed six operators, with a maximum firing rate of 2.5 rounds a minute or 0.75 when sustained. It needed 5 minutes to set up the battery, which could be reduced to four for departure. The ballistics use 100-kg HE (High Explosive) projectiles that can reach 25 miles with ERFB projectiles and 30 miles with ERFB-BB type, thanks to the 63-liter capacity of the barrel chamber.

Although it is extremely versatile, the option of purchase was discarded and self-propelled M-110 weapons were incorporated. However, development did continue in China, although there is neither news of that

HIGH OPERATIVE CAPACITY

Very robust, with a very modern and efficient design, and capable of adjusting its barrel length, this 155-millimeter weapon developed by STECSA ranks alongside the best weapons of this category.

MOBILITY TESTA

The SB-155/39 howitzer developed by the Santa Bárbara company includes an auxiliary engine at the front, which means that it can make small changes to its position without needing to be towed.

country's army adopting it nor evidence of any export contracts having been signed.

Santa Bárbara comes onto the scene

Towards the end of the seventies work began at the San Carlos Artillery Factory on the design work and finishing touches for a prototype gun that would eventually be called C.155/39 SC-80 REMA (Remolcado con Motor Auxiliar – Towed with Auxiliary Engine). From 1981, E.N. Santa Bárbara took over development and renamed it SB-155/39. This company expected to produce the towed version, with an APU and self-propulsion using a caterpillar chassis.

The weapon: ready for testing

After making a few last minute adjustments, the gun carriage was ready at the beginning of that decade, weighing 9 tons, incorporating a 39 caliber barrel with constant 48 rifled grooves, capable of sending projectiles 15 miles, measuring 12 meters in its operational position and 9.20 for being transported, with a height of 2.4 m and a weight of 2.5 m. It can fire in a vertical sector that ranges from –3° to +70°.

Testing showed that the design of the activating mechanisms was less than satisfactory, and its maximum range did not measure up to the demands of the foreseeable future. In general, this weapon did not live up to the requirements of the Spanish Ground Forces, although, at the same time, its firing was sufficiently accurate.

Solving the problems

The economic situation and doubts over which model to adopt (Santa Bárbara's or SITECSA's) considerably delayed the purchasing process of a 155 mm howitzer. So much time, in fact, that the latter company closed its design office and its production plants. Taking advantage of the shape of the gun carriage, wheels, APU

SHARED EXPERIENCE

SITECSA used two small workshops for making its range of guns and howitzers, ideal production sites for creating the artillery weapons that were due to be tested as prototypes.

EVALUATION OF THE PROTOTYPES

The Spanish Ejército de Tierra evaluated the prototypes of the 203-millimeter SITECSA gun, but rejected them in favor of similar M-110A2 type weapons with self-propulsion.

facility and production system developed by SITECSA, Santa Bárbara began work in 1995 on a more modern and versatile version. It incorporated a 52 caliber barrel, a standard that, it has to be said, had already been used on all other similar artillery weapons for several years. The development criteria were established in 1996, and the list of technical features were complete by 1997, and so production work on the 155/52 APU SBT-1 howitzer could begin. It received a barrel made by the German Rheinmetall company.

The Trubia factory was chosen as the location for the development of the prototype, which proved to be robust, manageable and highly mobile. The five main sections are the cradle and the elastic organs, the lower gun carriage, the upper gun carriage, the recoil mass and the APU. After the initial trials, it was unveiled to the press in May 1998, and exhibited at the EuroSatory show in Paris the next month, where it attracted a lot of interest. Its most outstanding characteristics include its range of 18,400 meters with standard M-107 projectiles and up to 25 miles with extended range or "base bleed" ones; it can be towed at a maximum speed of 55 mph and can move independently at 10 mph for positional changes. It has been fitted with a three stage cylindrical muzzle that reduces recoil by 35 %.

Results up until now have been encouraging, and there is talk of modifying its fastener with hydraulics and incorporating an automatic breech and a revolver that add new dimensions to the project, which is currently producing the initial series of six units.

Therefore, four could be developed for to the Ground Forces for evaluation before undertaking the internal order of over one hundred howitzers, to be shared between the Coastal and Field Artilleries.

SHUTTER DESIGNED FOR ARTILLERY

The design of the breech, the projectile fastener and the propulsive charge of the 155 millimeter SITECSA howitzer took into account the need to help with the work of artillerymen and to attain the fastest firing rate possible.

105/30 MM LG1
Despite its excellent characteristics, which have led to one hundred sales in four countries, the French army has not bought the 105/30 mm LG1 model, which would seem to be a good alternative for those armies that need a light, modern weapon.

The French policy of maintaining self-sufficiency in many aspects of its defense and the decision to maintain its Armed Forces relatively independent of western peace organizations inspired a special industrial development program with highly qualified and capable companies. This group includes GIAT industries, who have been working for several years on the production of a varied range of towed and self-propelled artillery systems that satisfy the demands for light and heavy weapons of the French Armée de Terre and a notable group of traditional buyers of French-made material.

Towed range of light and heavy weapons

The first modern designs of towed French artillery were the mod. 50 155 millimeter howitzer, which was completed just after the Second World War with the aim of providing a long range weapon that could hit targets over 12 miles away with standard munitions. Its many good qualities led to it being sold in high numbers to Ecuador, Lebanon, Sweden, Switzerland and Tunisia. Israel was added to the list, using it on an M4 Sherman tank caterpillar chassis and some towed mounts that were captured from the Lebanese after the 1982 invasion.

HIGHLY DEVELOPED TECHNOLOGY
This picture shows the tactical use of a group of Caesars working alongside advanced reconnaissance technology. They are linked by a control and command system that can fire accurately with conventional and intelligent munitions.

A weapon with superior features

The French Motorized Infantry and the Rapid Reaction Force requested, midway through the nineteen eighties, the development of a new, lighter, more modern and more powerful 155 mm weapon. The first six

prototypes were ready in 1987 and these were used for testing, and results were so favorable that by 1989 one hundred and five 155 TR weapons had been ordered. They were produced so quickly that some of them were even ready for use in the 1991 battles that liberated Kuwait from the Iraqi invasion.

Its characteristics (many of which were based on orders from Cyprus and another unconfirmed country) include a double support, which makes it highly stable during fire; a 39 caliber barrel with a double muzzle that can launch conventional projectiles over 15 miles, or 18 miles if they are rocket assisted; the installation at the front of the gun carriage of a small 39 hp engine and a driving position that means it can move independently at a speed of 5 mph and get across three-foot ditches.

It weighs 10.75 tons and measures 10 meters in the firing position. It can fire all NATO recognized projectiles. To satisfy other possible orders a prototype was developed with a 45 caliber barrel, and was followed by another of 52 caliber, which is the one that is attracting most commercial interest nowadays.

Other artillery options

For those countries that simply cannot afford to buy such modern artillery weapons, and use United States M114 155/23 mm models, GIAT offers a transformation pack known as M114F. The prototype was completed in 1990 and includes a French 155/39 TR barrel that can fire up to 5 times a minute, a crane that helps with loading operations, and

DETAIL OF A 155 MM WEAPON
This picture shows the layout of some parts of the French 155-mm TR. We can observe the loading assistance system on the right, the shutter in the center and, over on the left, the artilleryman's seat from where he controls the pointing system.

READY IN FIFTEEN MONTHS
It was 15 months from the day that the Thai army ordered 24 LG1 Mk 11 guns to the moment that the first 4 arrived in Thailand, just in time to take part in a firing exercise that took place in February 1997.

modified pointing elements that reduce the number of people needed to fire the weapon from 11 to 7. GIAT offers a similar option for the 105-mm M101A1 weapon. Thailand ordered three hundred of these conversion packs.

Even more modern is the Caesar (CAmión Equipée d'une Sistéme d'ARtillerie). The original idea was to position a piece of 155/52-mm artillery on the trailer of a 6x6 truck, which would reduce the overall weight. This would give the system many tactical advantages due to the potential 350 miles range and a maximum road-speed of 55 mph. At the same time, it would provide protection for the six artillerymen that would travel in the armored cabin. This also gives the advantage of being able to leave the battery in less than a minute, firing three shots in just 15 seconds, thanks to the provision of 18 transported munitions on board, and it could reach targets up to a maximum of 26 miles away.

Light with a long range

These qualities define the LG1 gun carriage artillery of 105 mm proposed by GIAT. The first prototypes were developed in 1987, inspired by the French company's proposal (dating from early 1986) of increasing its potential within a market that was also offering similar weapons

produced in Great Britain. A model was developed that resulted in contracts being signed for 28 destined for Canada, 20 for Indonesia, 37 for Singapore and 24 for Thailand. At the moment, a favorable decision from the French army is still being awaited, but if it were to happen, they should order around a hundred units.

Its most outstanding characteristics include the fact that it only weighs 1.5 tons, meaning it can even be towed by light four-wheel drive vehicles such as the Peugeot P4 or a Land Rover and its 30 caliber barrel includes a conventional breech for firing all United States M1 type munitions up to distances of 12 miles. The LG1 Mk II variant has been updated to use higher-pressure munitions. It can enter battery in just 30 seconds thanks to the hydraulic system that activates the base plate situated beneath the gun carriage and between the wheels, allowing it to be towed at high speeds.

Self-propelled weapons for internal and export use

At the beginning of the nineteen fifties the Aletier de Construction at Tarbes and at Roanne completed a self-propelled weapon that included a 155 mm gun and derived from the towed mod.50 with the chassis of the light AMX-13 tank. This model came to be called the Mk F3 155/33 self-propelled howitzer.

The mobility of the weapon (and not the protection, for it is an uncovered mount without an armored casemate) was what primarily inspired the production of 222 units for France and several more, about 400, to be exported to Argentina, Chile, Cyprus, Ecuador, Kuwait, Morocco, Qatar and Venezuela.

In search of better features

Despite the fact that the Mk F3 was capable of support fire at distances of up to 15 miles when it used "base bleed" munitions, its radius of effect was 280 miles for the version propelled by a diesel engine of 280 HP and it only needed two crewmembers to operate it. Its poorer features with respect to artillery and the slowness of its loading capacity compared with other systems on the market meant that a new version needed designing.

This was called 155 GCT (Grenade Cadence de Tir), went into development in 1969 as

MOBILE ARTILLERY ON A TRUCK

The Caesar is a new concept in artillery, with the installation of a 155/52-millimeter weapon on the trailer of a truck with an armored cabin for transporting the operators of the weapon and the 18 munitions for immediate use.

FUTURISTIC CONCEPT

The Caesar system is an innovative artillery unit that includes the installation of a 155/52 mm weapon on the back of a medium light tactical truck, thus making it more mobile and protecting its crew (left photograph).

ADVANCED MUNITIONS

GIAT includes in its product range the development of advanced munitions that seek to satisfy varying requirements. These include the three aerodynamic 155-mm projectiles shown here (right photograph).

the result of an order from the French army, who wanted to replace its older models. The first prototype was finished in 1972 and was unveiled to the public at the 1973 edition of the Satory Exhibition, near Paris.

The first pre-production units were completed between 1974 and 1975, and they were then subjected to evaluation tests. Its armor-plated casemate, fitted with an automatic loading system, and its articulated chassis, that allowed the piece to move across all types of surface and that was derived from that used by the medium AMX-30 combat tank, generated a lot of interest.

In 1992, a version was unveiled that would be destined for export. This used the same turret, but it was now installed on the chassis of a T-72 combat tank. At the same time a different version fitted with a 52-caliber long gun that notable increased its range was launched.

High artillery capacity

In 1977, the first ones were produced (of a total of 400 made to date) for the Saudi Arabian army, which became the first country to have 51 of these weapons. The French army selected it in 1979 and has so far received a total of 179 units of the AUF1 155/40-mm variant and 74 of an updated version known as AUF1 T.

Iraq acquired 85 and Kuwait, 18, but it is unlikely that many of these are still active as a result of the Gulf War.

Basically, this weapon consists of a medium tank chassis onto which a gun turret has been installed to give protection and support to the

TECHNICAL CHARACTERISTICS LIGHT TOWED LG1 MKII HOWITZER

COST IN DOLLARS:	**430,000**	**WEIGHTS:**	
CALIBER:	**105 mm**	Total in combat order	1,520 kg
DIMENSIONS:		**FEATURES:**	
Length in firing position	6.6 m	Range with minimum load	0.8 miles
Length in transport position	5.32 m	Range with maximum load	10 miles
Height in transport position	1.34 m	Angle of elevation/depression	+70°/3°
Width	1.97 m	Firing rate	12 rounds a minute
Height off the ground	0.3 m	**CREW:**	**7 men**

155-mm artillery weapon that can destroy targets at 18 miles thanks to its long barrel. This model is particularly renowned for the light materials that it is made of, which reduce two tons from the weight of the chassis of the original version. The commander and the artilleryman sit on the right-hand section of the turret and the loader on the left. At the front of the turret are some vertical loaders configured for six circular gun carriages and which provide the total capacity for 42 rounds that can be fired at a rate of 8 rounds per minute when using the automatic system, or three rounds per minute when reloading is carried out manually.

Another significant factor is its high level of self-defense, owing to a medium or heavy anti-aircraft machine gun and the twin smoke launchers which are built into their turrets. It only takes two minutes to prepare the machine for combat, one minute to leave the firing position and fifteen for the fifteen men to reload the munitions and propulsive charges. The integrated fire control system is made up of an optic goniometer, the CITA 20 system with built-in navigation and a sight that allows for the use of the weapon in heavy fire for self-defense against important targets.

CLASSIC CONCEPTION

The light French LG1 MK II includes a 30 caliber barrel fitted with a chamber with the ideal volume for using M1 type munitions (with a 10 mile range) that are standard all over the world.

40 CALIBER BARREL

The French made self-propelled GCT howitzers are famed for their turret, which includes a 155-mm diameter and 40 caliber long barrel. At the back, it has an automatic loading system that means it can fire eight times in just one minute.

The need for more mobile artillery units during operations so as to vary locations and thus avoid enemy anti-battery fire inspired military designers to configure a mobile chassis and a mount situated in the upper section. The evolution of this concept influenced, nearly a quarter of a century ago, the creation of a family of United States 155 millimeter M109 howitzers. Some five thousand of these artillery units have been exported, and an even higher number were built for use by the United States Armed Forces.

Washington Conference

In January 1952 an important conference was held in Washington D.C. to discuss the urgent need for a self-propelled artillery weapon that would be highly capable with respect to both mobility and firepower. The first studies concerning what was then called the Howitzer 156 mm Self-Propelled T196, were presented in August of the same year. The concept was finally approved, despite the first set of proposals being rejected in 1954.

The process of concretion begins

The military leaders' decision to unite the development of the T196 to that of the 110 mm T195 was given the go-ahead and, in June 1956, it was decided that the former should be fitted with a 155 mm weapon and the latter with a 105 mm one. The final adjustments were made to the scale model in October 1956, and the construction of the first prototype was authorized. It ended in 1959 and was moved on to Fort Knox for preliminary evaluations, in which a few significant faults were found relating to the suspension.

AMERICAN DESIGN

The M109 is a highly exported US design. More than fifty thousand units have been built for the US Army and US Marines and they have been employed in the brigade and division artillery groups.

PREPARING A FIRING ACTION

Before starting to fire M109 howitzers, the projectiles that will be fired must be prepared, an action which requires the positioning of the fuses and the connection of the propulsion charges that are needed for the action.

At the beginning of 1961, two T196E1 were ordered, and these were incorporated into the diesel engine-propeller, which was classified for limited production in December of the same year. The process, supervised by the Cadillac Motor Division, commenced at the Army Tank Plant at Cleveland. The first production units were ready by October 1962 and from July 1963 onwards it was classified as M109. Up until 1969, 1,961 units were made for the Army and 150 for the Marines.

BMY started the production of the M109 in 1974, and until 1993 made 4,242 of the M109A1B and A2 types. These incorporated several improvements, in particular the new gun with ranges up to 18,100 meters (compa-

red to the 14,600 of the original model), a distance that, using projectiles with auxiliary propulsion, could reach 24,000 meters (nearly 15 miles).

The A3 model inspired the transformation of many A1, which received an M178 gun carriage and several minor improvements. The A4 incorporates a hydraulic power system and a filter system against nuclear, biological and chemical (NBC) agents, while the A5 comes as the result of introducing several changes that affect this artillery weapon with a larger breech, hydraulic systems, transmission and engine and wheel elements.

Orders arrive from all over the world

The features of the M109 system, unique to its genre at the time it appeared, attracted the interest of several countries in either buying American units or in making their own under license. One of the earliest was Germany, who made more than six hundred M109G, Italy, whose units were constructed by OTO Breda, and Egypt, which is installing a turret with a 122 mm piece onto the United States caterpillar chassis. Later, such important countries as Saudi Arabia, Austria, Canada, the United Arab Emirates, Spain, Ethiopia, Greece, Iran, Israel, Libya, Jordan, Kuwait, Morocco, Norway, Peru, Switzerland, Thailand and Taiwan were added to the list.

These and other nations have some 4,500 M109s of all versions except the A6, which is currently still at the promotion stage. A clear example of this is that the new model has recently been acquired by Kuwait to complete its upgrading program (started after the end of the Gulf War) that expects to incorporate a 45-caliber barrel onto them. To the earlier artillery weapons one would have to add the M992 FAASV munitions vehicles, which have a very similar configuration and tend to accompany the howitzers to increase the possibility of carrying out sustained fire for longer periods and in greater measure.

Developed for the next century

The need to confront new challenges with respect to the kind of artillery needs that armies foresee in the early decades of the new century have inspired several new upgrade programs that set new standards in the capacity, reliability and power of these self-prope-

LONG RANGE CAPABILITY

The inclusion of a 39 caliber barrel on many of the more modern M109s (or those that have been modernized) allows them the possibility of firing their projectiles at objectives that are 15 miles away, a distance that goes up to 18 miles if rocket assisted projectiles are used.

AMPHIBIAN WEAPONS

The Spanish Infanteria de Marina has 6 M109A2 howitzers supported by an M992 FAASV munition supplier, which greatly increases the rate and capacity of fire, thanks to the munition reserve that it transports.

lled weapons.

The United States Paladin

The first work on what was defined as the Howitzer Improvement Program (HIP) began in November 1984. Different companies' proposals were first studied four months later. In October 1985, the project presented by BMY was chosen.

Four prototypes for the United States Army and two for the Israeli Army began to be modified under a program that took three years. Eventually, on the 30th of March 1988, the upgraded howitzers were unveiled (known as M109A3E3) and they underwent intense tests at the centers in Yuma and Aberdeen. After getting both technical and operational approval (the latter at Fort Still), production was started for the first order for 44 units signed in September 1990. The upgrading cost one million dollars per unit.

Later renamed the M109A6 Paladin, these systems began to be presented from April 1992, and new orders were soon received. By October 1998, 630 had been completed. The FMC Corporation joined in with the work, and eventually merged with BMY Combat Systems under the new banner of United Defense LP. The modifications introduced included a new turret with improved armor and internal Kevlar lining; an automatic control system; ballistic processor and navigation system; a 155/39 mm M284 barrel with a range of 30,000 meters with the M203/203A1; better munitions stores; an air conditioning system and a long list of

GOOD MOBILITY

The configuration of the wheels, the power of the engine and the characteristics of the support vehicle allow the crews of M109 howitzers to travel large distances to get into position.

other modifications that improved its survival by 150 %, 100 % in response time, 30 % in range and 25 % in crew efficiency (which continued consisting of four men). It weighs 28,849 kg and has a total length of 9,804 meters.

The European programs

The delay in the design and manufacture of a system that could replace it led to the approval for the conversion of six hundred German howitzers, weapons that are currently known as M109A3GE and are renowned for their 39 caliber barrel and 18.8 liters capacity breech. This means they can stay active until at least the year 2005. As for the Italians at OTO Melara, they worked on converting several batches of howitzers, which were to be called M109L and whose barrels were compatible with the FH-70 towed model. The Swiss are updating theirs to the standard proposed by Thun, who suggest a 47-caliber barrel and an additional magazine to be installed in the rear section of the turret. The Taiwanese, meanwhile, have replaced the original turret with a towed piece installed in an aluminum structure and thus creating what is known as the XT-69.

RAPID FIRE ACTIONS

Four or six M109 howitzers, depending on the country and model, make up each firing battery. These must work with speed and accuracy so as to quickly move to a different location after firing and therefore escape the threat of enemy fire.

In Spain (despite the continued cutbacks in the defense budgets), 72 howitzers are being converted into the A4+ and A5 variants. A large part of this process is being carried out at the Centro de Mantenimiento de Sistemas Acorazados (CMSA) number two in Segovia.

SMALL CREW

Four men are plenty to operate the different elements of a self-propelled M109A5 artillery weapon. These might include the commander, the aimer, the loader and the driver, who also helps out with the handling of projectiles.

MUZZLE

An effective muzzle has been fitted at the front of the 155 millimeter barrel, with side chambers that reduce both the recoil caused by firing and the effects of the fire on the vehicle and crew.

RECUPERATING ELEMENT

The artillery gun carriage is visible at the front, fixed to the main barrel. It acts as a recuperating element during the recoil provoked by the shot and forms part of the elevation mechanism that aims the weapon.

INTERIOR OF THE TURRET

Inside the tower there are several pieces of navigation and aiming control equipment, the enormous shutter that forms part of the interior of the barrel, some of the buffer mechanisms, the user positions and the magazines.

BARREL CONNECTION

More modern howitzers, as well as those that have been modernized into more capable versions, include an element at the front of the vehicle that is used for connecting the main barrel automatically, making it easier to move the weapon to different places.

TECHNICAL CHARACTERISTICS M109A5 SELF-PROPELLED HOWITZER

COST IN DOLLARS:	3,000,000
CALIBER:	155 mm
Length	9.804 m
Height	3.236 m
Width	3.149 m
Height off ground	0.457 m
Width of the track	0.381 m
WEIGHTS:	
Total in combat order	28,849 kg
PROPULSION:	
8V-71T LHR Detroit diesel turbine engine of 440 HP and 2,300 rpm	

FEATURES:	
Range with standard munitions	15 miles
Range with assisted munitions	18 miles
Angle of elevation/depression	+75º/-3º
Firing rate	4 rounds a minute
Maximum speed	40 mph
Autonomy	215 miles
Munition stores	39 155-mm projectiles and 500 of 12.70 mm
Complementary armaments	1 12.70 x 99 mm M2 heavy machine gun
CREW:	4 men

HEAVY MACHINE GUN

The self-defense capabilities of M109 howitzers is provided by a gun carriage that gives mobility and assists with the handling of a Browning M2 heavy machine gun (a very effective and powerful weapon) with a caliber of 12.70 x 99 mm.

SUPPORT FOR THE PROJECTILES

The magazines are kept in the turret, where the 155-millimeter projectiles that make up the weapon's immediate munition reserve can be stowed and transported.

THE REAR

Several containers and supports are kept at the back of the turret for transporting different parts related to the vehicle. The large frames that keep the weapon steady during firing are also stored here.

HIGHLY MOBILE

The wheels include a drive wheel at the front and seven road wheels that move the track and make it appropriate for any type of surface.

The need to face the 21st century with newly developed and more capable weapons adapted to the new technology and requirements of the armed forces, has led to the purchase of all kinds of new equipment. However, only the more economically powerful nations have been able to do this. The case of Germany is particularly significant. Despite the process of reunifying the former West and East Germanies (which is costing the Federal government enormous amounts of money), work has continued on all the Army programs initiated in earlier decades. These include the attack helicopter, new groups of combat vehicles, missile systems and artillery weapons. The most outstanding of the latter is the self-propelled PzH 2000 howitzer, which we shall be examining in this chapter.

OBJECTIVE: TO BE THE MOST CAPABLE

The design, features and capacity of the PzH 2000 make this the most capable of western howitzers. This can be proved by the interest that several countries have shown in acquiring it or in their own industries receiving the production license.

A long process of preparation

Since the late nineteen seventies, several countries, including Italy, the United Kingdom and Germany, have been working together on the development of a self-propelled howitzer (known as the 155/39 millimeter SP-70), which has inherited many of the trademarks of the 155 mm FH-70 towed gun (of which more than half a million were built).

The Panzerhaubitze 2000 is conceived

The cancellation of the international program in July 1986 caused the German government to seek the proposals of different industrial groups to make a weapon that the German army defined in the following way: with the capacity to transport 60 rounds and their corresponding charges, the installation of an automatic loading system that would greatly improve its firing rate, a range of 18 miles with unassisted projectiles, the provision for a team that could operate it in areas seriously contaminated by nuclear, biological or chemical (NBC) elements, excellent mobility, capable of operating independently and armored sufficiently well to withstand attacks from enemy weapons.

PANZERHAUBITZE 2000

Designed in a relatively short time, the German PzH howitzer is known for its excellent tactical mobility and the accuracy and speed of its artillery fire.

Development work begins

The design and development work on what was then known as SP2000 was assigned to two industrial consortiums that were contracted to carry out their two respective proposals. This phase required the investment of 183 million German marks. The first of these was made up of Krauss-Maffei, KUKA and Rheinmetall, and the second was formed by Wegmann & Co. and Mak System Gesellschaft, and both began work on their projects in October 1987, when Phase One was given the official go-ahead.

During this phase, each of the two consortiums worked on constructing a prototype that integrated a 39-caliber weapon. At the beginning of 1988, it was decided that a 52 caliber one was preferable, which would improve the range and operative possibilities of the weapon, an element that was designed and manufactured by Rheinmetall. One barrel was retained by the manufacturer and the other two were given to the two consortiums, who continued the work. The first group called their design KM, while the second group called their own project WECO. To provide mobility the chassis of modified Leopard 2 and Leopard 1 combat tanks were used respectively.

Intense developments

After completing the design process, the prototype pieces were completed and presented to the Army for evaluation at the end of 1989. Nearly a year was spent testing and comparing the two to decide which would be the more appropriate one to adopt. At the end of 1990, it was decided that the WECIO project was the most suited to the armies' requirements. From there, Phase Two of development was proposed to both companies (costing 195 million German Marks), requesting that they develop four more howitzers for a wide variety of tests. Then, production agreements were signed for what eventually came to be called PzH 2000, and was ready in the mid nineties.

WHEELS

The design of the wheels and the provision of an engine-propelled group at the front give the PzH 2000 excellent capabilities with respect to its ability to move across any type of surface.

AUTOMATIC LOADING SYSTEM

The automatic loading system and its projection charges were designed to improve the firing rate that would be possible with just manual fire.

In the fall of 1992 the first of these prototypes was ready, with the others coming after one month intervals. The original weapon was also modified to this new standard. Technical evaluation began in 1993 and took more than two years. Over this period of time, the weapons underwent tests of their potential, the study of their logistic needs and other important aspects. At the same time, the original request for 1,254 units was reduced to 594, which is more in line with current requirements.

The first units, part of an initial order for 185, were completed and delivered on the 1st of July, 1999, and the remainder are expected to be ready by the year 2002, although export orders have already arrived, and therefore the production line will probably remain operative for some time after that. The first of these orders was Italian, where a license agreement for the Italian construction of 70 PzH 2000 has been signed. These units will be distributed among the self-propelled artillery regiments of the

Ariete, Garibaldi and Centauro brigades. They will include batteries of nine howitzers, including one moving section, one traveling section and one firing. Another possible buyer is the Spanish Army (which has expressed an interest in fifty of these howitzers to complement its Leopard 2 tanks), along with Denmark, Norway and Sweden.

Tomorrow's technology available today

This enormous self-propelled howitzer possesses several unique qualities that make it the most advanced example of its type anywhere in the world today. The most important fact is that it only takes 30 seconds to stop and open fire, whereupon it can fire 10 rounds in a minute, and then in another 30 seconds its is already moving on to the next firing position.

Upgraded artillery and feeding system

The PzH 2000 weapon is an L52 piece developed to complement NATO's Joint Ballistics Memorandum of Understanding. Its barrel has a 155-millimeter inner bore and has been covered in chrome to reduce wear.

SELF-PROPELLED ARTILLERY

The length of the barrel of this new self-propelled howitzer from Germany was designed to reach distances of between 18 and 25 miles. This is enough to defeat an enemy from a safe distance. The latest firing rate is around 10 shots a minute.

CROSS SECTION

The vehicle for the PzH 2000 howitzer was based on the Leopard 1 tank, although there is a more rational layout of elements and more space for the artillery weapon.

It has an effective muzzle that considerably reduces recoil and different sensors, such as those for temperature and the speed of projectiles as they take to the air, in the breech. With this (whose angles of operation lie between –2.5° in depression and +65° in eleva-

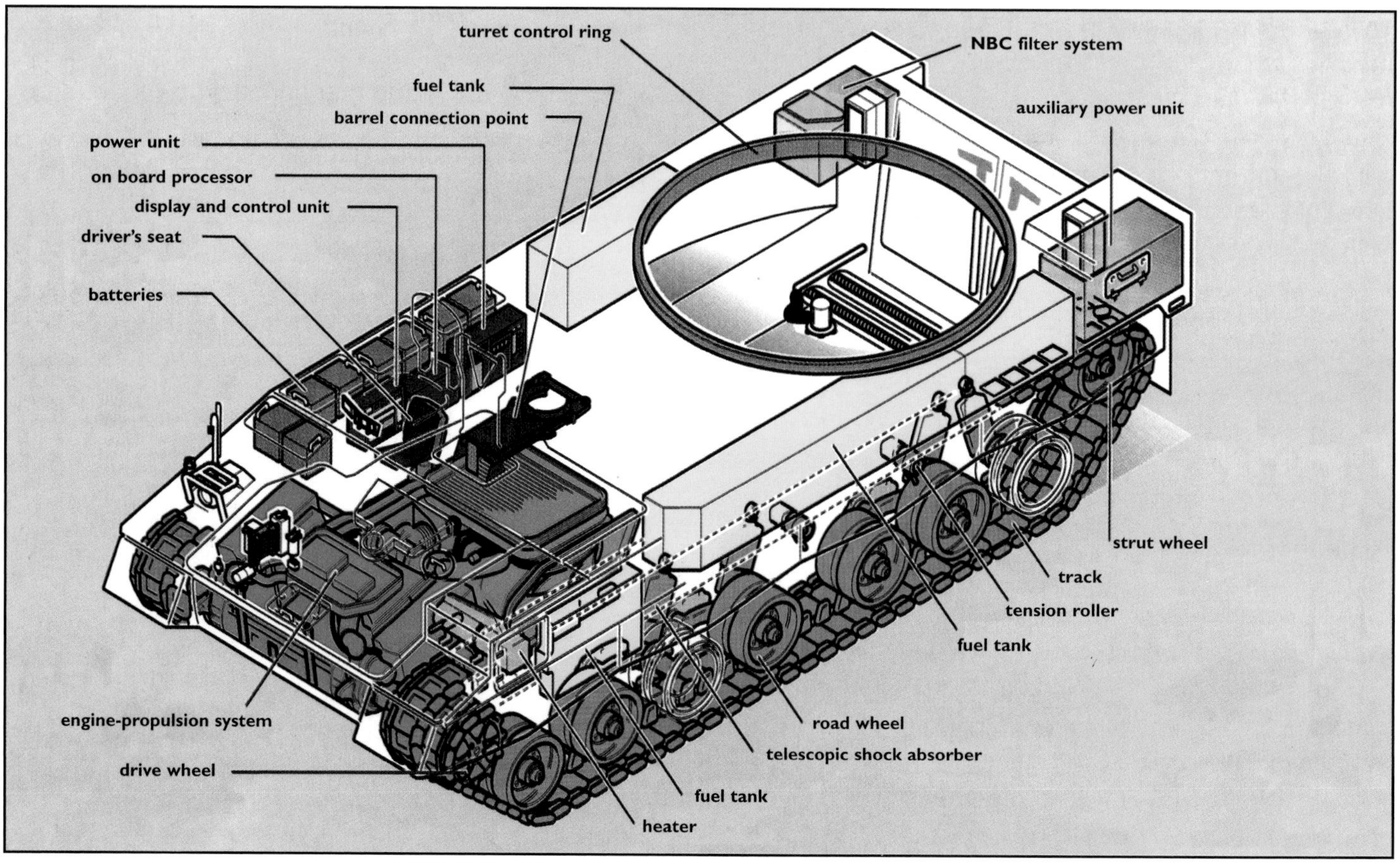

TECHNICAL CHARACTERISTICS SELF-PROPELLED PZH HOWITZER

COST IN DOLLARS:	6,000,000
CALIBER:	155 mm
DIMENSIONS:	
Length	11.67 m
Height	3.43 m
Width	3.58 m
Height off ground	0.55 m
Width of the track	0.55 m
WEIGHTS:	
Total in combat order	55,000 kg
PROPULSION:	
1,000 HP MTU 881diesel engine	

FEATURES:	
Range with standard munitions	18 miles
Range with assisted munitions	25 miles
Angle of elevation/depression	+65°/-2.5°
Firing rate	3 rounds in 10 seconds and another 10 in a minute
Maximum speed	40 mph
Autonomy	280 miles
Munition reserves	60 howitzers and 288 modular propulsion charges
COMPLEMENTARY WEAPONS:	
One 7.62-mm MG-42 medium machine gun and 8 dual purpose grenades	
CREW:	
5 men, although it can be operated with only 3	

tion), ranges of over 18 miles can be reached when using the very latest in standard munitions, or 25 miles using rocket-assisted types. There is no similar howitzer that could achieve such distances, and not only that, it also has excellent self-defense features against enemy anti-tank activity. Related with the former is the automatic munition-loading, composed of an electro-pneumatic system that moves the projectiles and their charges from their stores-position to the breech in a fast and effective way. By doing this, up to three rounds can be fired in just ten seconds, followed by another nine or ten in less than a minute, with the fuses adjusting themselves automatically. Twenty rounds can be fired in two and a half minutes and up to sixty rounds can be fired in twenty minutes.

It is also worth mentioning that it can be loaded and unloaded in a semi-automatic mode, or if conditions require, completely manually. It only takes two men a time of 11 minutes to put the 3.4 tons that the 60 projectiles and their propulsion charges weigh onto the howitzer. Its self-defense facility is a medium 7.62 x 51 mm MG-42 machine gun and 8 smoke and anti-human grenades.

Highly mobile and comfortable for the crew

The chassis of this model was inspired by that of the Leopard I tank, although it has been modified with a longer set of wheels, an new propulsion group at the front (based on the MYU 881 diesel engine of 1,000 HP and that provides a ratio of 18 HP/t), a redesigned rear area that gives access to the transport zone and a well-designed driving position with a multi-purpose display panel that receives messages and warnings. The auxiliary power equipment and the NBC filters are positioned in two crates situated at the rear. The set of wheels includes three integrated fuel tanks on the side of the vehicle, which provide the capacity for a range of over 280 miles. Its maximum speed is about 40 mph and it is able to cross 3 and 1.5 meter ditches

OPTIMIZED PROJECTILES

The Rheinmetall W&M company is developing new artillery projectiles, among which is this optimized model with a Global Positioning System (GPS) that can achieve levels of accuracy that are far superior to those of conventional projectiles.

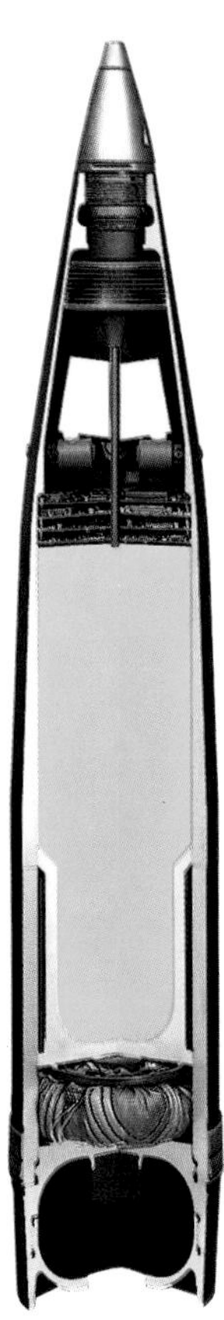

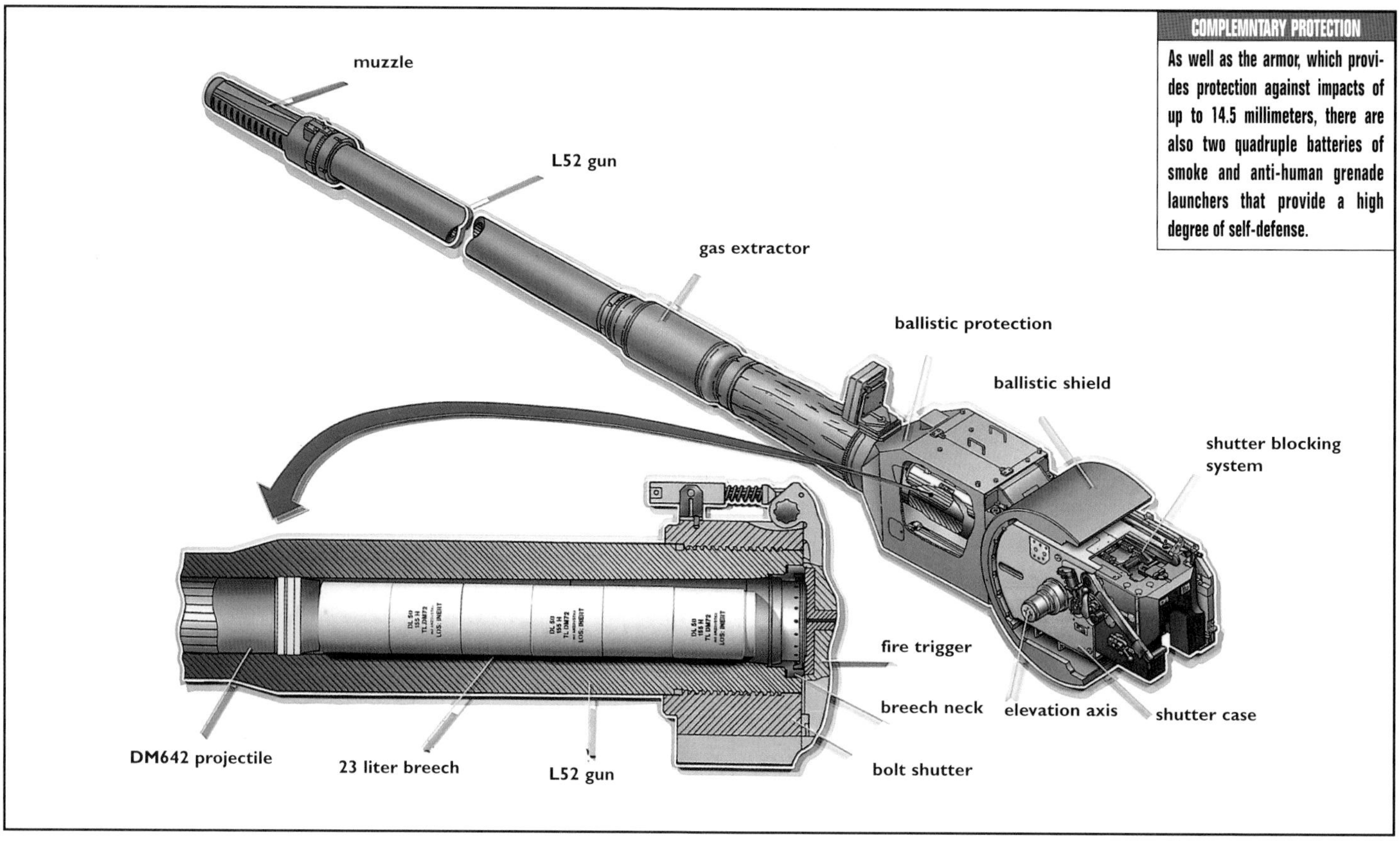

muzzle
L52 gun
gas extractor
ballistic protection
ballistic shield
shutter blocking system
fire trigger
breech neck
elevation axis
shutter case
bolt shutter
DM642 projectile
23 liter breech
L52 gun

COMPLEMNTARY PROTECTION

As well as the armor, which provides protection against impacts of up to 14.5 millimeters, there are also two quadruple batteries of smoke and anti-human grenade launchers that provide a high degree of self-defense.

The howitzer is fitted with basic armor plating on the inside, using a spall liner type material, which can withstand the impact of projectiles of 14.5 mm, grenade and howitzer splinters and neutron radiations. A ceramic paneling system can be incorporated into the upper section of the turret to withstand the effects of anti-tank weapons. The crew itself is protected by an NBC filter and ventilation system and an automatic fire extinguishing system. The walls of the magazine were designed so as to explode outwardly, and there are two large emergency doors at the back.

Whilst on the subject of these protective measures, it is also worth mentioning the work ergonomics, derived from the installation of a fire and systems control system that operates via display screens that facilitate the tasks of the artillerymen and assist the powers of communication between man and machine. These include a fire control processor for ballistic calculations, the electrical system that permits complete control of the main weapon and an autonomous navigation system. The commander also has a panoramic periscope for day or nighttime use, which incorporates a laser measure that can be used for firing under heavy fire against targets that represent an immediate threat to the howitzer.

NEW MODULAR CHARGE

The DM 72 155-millimeter modular type charge was designed to maximize the capacity of the 23-liter breech of the L52 gun. Here we can see some details of its configuration and constitutive elements.

LONG RANGE HEAVY ROCKET
The MLRS rocket launcher can fire both modules of six rockets and those that contain ATACMS, capable of sending a thousand submunitions 80 miles. A new version is currently being developed that would reach 190 miles.

Rockets have been used for military purposes for several centuries, although they have only really become common since the Soviet Katiuska were so massively used in the Second World War. Of the most modern systems now used in the West, there are few so powerful as the United States Multiple Launch Rocket System, known all over the world as MLRS. These are self-propelled modules that can destroy all kinds of targets with their rockets. The Iraqis know this only too well, having suffered their devastating effects in the early phase, before the air and ground attacks, of the Gulf War in early 1991 that sought to liberate Kuwait.

SEMI-AUTOMATED CONTROL
This rocket artillery system can carry out all of its functions with the help of just three men who are responsible for loading and unloading the containers, lining the launcher up to its target and deciding upon the rate of rocket fire.

Created in the USA with an international outlook

The US Army made it clear that it required an artillery saturation system, and so the Redstone Arsenal in Alabama (part of the United States Missile Command) initiated, in early 1976, the preparatory phase for the idea of low-cost but highly effective rocket systems. Known as the General Support Rocket System (GSRS), this project investigated the development of their conventional rockets, providing them with a better rate of fire that would allow them to defeat both troops and light equipment, aerial defense systems and command centers.

The studies investigate new concepts

The Boeing, Emerson Electric, Martin Marietta, Northrop and Vought companies signed a contract in 1976 through which they received government subsidies to start preparing their respective proposals and to carry out the first work on this new design. In September 1977, the Boeing Aerospace and Vought Corporation proposals were selected.

Both signed a contract to make three prototypes of the Self- Propelled Launcher-Loader (referred to as SPLL) and the associated rockets. They were given 29 months to complete the task. In 1978, while the finishing touches were being put to these proposals, NATO showed (after the USA had invited them to study their progress) interest in this weapon, which could also be manufactured in Europe, where the system was renamed MLRS.

After the evaluation tests had been carried out at White Sands Missile Range in New Mexico, the Vought design was finally chosen, although the company itself had since changed

their name to Loral Vought Systems Corporation. At their Dallas factory, in Texas, the manufacturing process got under way, assisted by other United States firms, such as the Atlantic Research Corporation, that worked on the solid engine fuel, the Bendix Guidance Systems Division, that worked on the reference and stabilization system, the Brunswick Corporation, that dealt with the launch tubes, Norden Systems, who made the fire control system, and Vickers, who were put in charge of the driving position. Different production tasks were carried out in Europe by different companies based in Germany, France, Great Britain and Italy, where it had been decided that it should be adopted and integrated into the MLRS-Europaische Produktions Gessellschaft GmbH consortium, who set up their headquarters in the German city of Munich.

The first units are presented

In 1983, the artillery units of the US Army received the first definitive self-propelled units on the M270 armored launcher that had been built based on a modification and extension of an M2 Bradley combat vehicle. The first to be built in Europe were completed from 1989 onwards at a rate of ten units a month.

From then on, about 900 have been built for the United States, who have ordered more than 700,000 tactical rockets, and deployed 230 of them as part of the Desert Storm operation, where their extraordinary destructive power was proved.

Other countries that have started using this system are Germany, who have 156 launchers spread between eight artillery battalions; Bahrain, which bought 9 in 1992; Korea, who bought

MODULAR DESIGN

The MLRS was designed to work with modules that have the capacity for six 227-millimeter rockets that are automatically placed in the launcher. This can be substituted by a module for the larger ATACMS rocket.

MADE IN GERMANY

The German Ground Forces can deploy 8 battalions of MLRS systems, of which they own a total of 156 units, a versatile, powerful and reliable artillery system that can deal with advancing enemy forces.

29 in 1997; Denmark, with 8; France, who use 55 that they finished receiving in 1995; Great Britain, who received a total of 63; Greece, who ordered 18 launchers in 1994; Holland, who would become the first European country to deploy its 22 MLRS; Israel, who first asked for 6 and then a further 42; Italy, who has 22; Japan, now making 50 launchers under license; Norway, that has already acquired 12 and Turkey, who operate 12 systems and are still waiting for 24 more.

There are several other countries that have shown interest, the most important being Spain, Malaysia, Sweden and Switzerland.

An advanced and effective concept

Generally speaking, the MLRS is a caterpillar vehicle with six wheels and a cabin at the front (where the driver travels along with two weapon operators with armored protection from the blast of the rocket when it is launched and from the impact of light enemy fire), and at the rear is an erection element in two sextuple rocket modules.

Modular concept applied to different versions

Each of the above serves for both transporting and storing the rockets as well as launching them, and weighs 2,304 kilograms when fully loaded. They include an outer structure of aluminum that is fixed to six fiberglass tubes.

Each one of these can be automatically unloaded by the launcher itself, which includes the necessary elements for carrying out the module reloading operation autonomously. This is generally carried in tactical trucks.

HIGHLY CAPABLE FIRE

The unguided artillery system of MLRS has extraordinary firepower, being able to fire its 12 rockets in less than a minute against any objective within 18 miles, where a thousand submunitions capable of penetrating light armor will fall.

The control of the number of rockets to be fired, the angle of launch and other parameters of the fire are undertaken from the fire control system (FCS). An improved variant is already being introduced with more capacity that reduces response time from 5 to 1.5 minutes.

Different kinds of rocket can be fired by the system, including the M77 loaded with 644 submunitions, the AT2 with disseminating elements for 28 antitank mines, the M28A1 for launching in less open spaces, the ATACMS, that transports 950 M74 antipersonnel and material bombs over distances over 80 miles (only one fits in each launch module), and a version of the previous one that mobilizes 13 BAT self-guided submunitions. One has also been tested that has a Global Positioning System (GPS), which would ensure accuracy of less than two meters.

WIDELY USED IN THE USA

The US Army and National Guard have some 900 multiple rocket launchers that make up the backbone of their artillery force. They can be deployed wherever they might be needed with the support of their fleets of transport aircraft and ships.

THREE CREWMEMBERS

The commander, the launcher and the driver are the three crewmembers of an MLRS system, and have special seats in the cabin, where the fire control display screens and the communication systems inform them of the instructions related to their movements and fire actions.

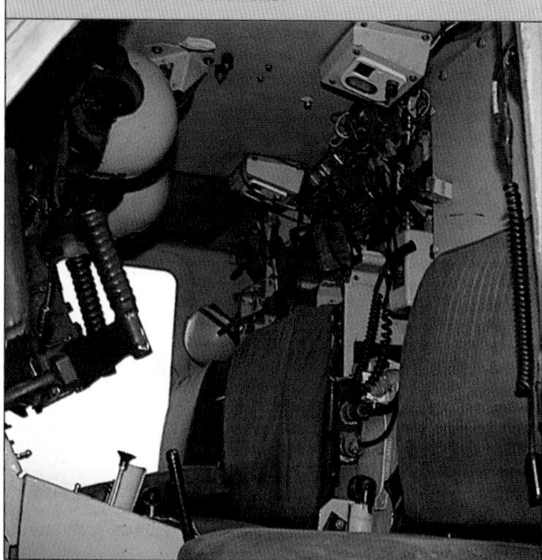

INTEGRATED COMMUNICATIONS

The MLRS integrates the necessary communication equipment for keeping contact with the command center or receiving radio information concerning the targets to be attacked, systems that take advantage of the capacity of the two antennas on the upper section.

ARMORED CABIN

This frontal picture of an MLRS launcher shows the positions of the three observation windows protected by armored blinds that prevent the rocket blast from hurting the three operators sat within, and also serve as protection from light fire.

HIGHLY MOBILE ON ALL SURFACES

The chassis of the MLRS derives from that used by the armored Bradley caterpillar with respect to design and theory. It incorporates a traction wheel at the front, a tension wheel at the back, and six smaller road wheels that help move the track.

PROPULSION GROUP

It is here that the Cummins VTA-903 turbo diesel engine is situated, producing 500 HP at 2,400 rpm. It drives the vehicle at a maximum speed of 40 miles an hour.

TECHNICAL CHARACTERISTICS MLRS ROCKET SYSTEM

COST IN DOLLARS:	5,000,000 for the launcher and 12,000,000 for rockets
CALIBER:	227 mm
DIMENSIONS:	
Length	6.972 m
Height	2.612 in normal position and 5.925 with the launcher raised
Width	2.972 m
Height off ground	0.43 m
WEIGHTS:	
Total in combat order	25,191 kg
Of a standard container	2,300 kg

PROPULSION:	
Cummins VTA-903 turbo diesel engine producing 500 HP at 2,400 rpm	
FEATURES:	
Maximum speed	40 mph
Autonomy	310 miles
Mobility	60 % in inclination and 40 % laterally
Range of M77 rocket	18 miles
Range of ACTAMS rocket	70 miles
Range of ACTAMS Block 1ª rocket	180 miles
Rate of fire	12 M77 rockets in less than a minute
CREW:	3 men

LAUNCHER ERECTOR

A metal structure that can turn left and right is what gives support to the missile containers and includes hydraulic elements that establish the optimum angle of inclination that the range of fire needs.

GAS EXHAUST

At the back of the MLRS are 12 tubes that contain the rockets. When they are activated, they generate a great deal of gas and a severe blast at this section of the system.

SIGNALING AND BRAKING ELEMENTS

The positional lights and brakes are situated at the rear along with hooks for pulling the vehicle whenever it might get stuck in the ground and a central connection for pulling towable weapons and vehicles with wheels.

The need to counter the threat to ground forces created by different types of aircraft led to the introduction of small-caliber artillery systems that were later to be replaced by short-range anti-air missile launchers. The first generations of these missiles included a homing device that was guided by the heat source generated by the exhausts from the turbines of airplanes and helicopters. This way it was directed towards them, and exploded on impact due to its explosive charge, although this did not always happen with the accuracy that both the manufacturers and their clients had expected.

EASY TO OPERATE

Just one man can operate this light launcher without problems, by locking the sight onto the target aircraft, firing the rocket and reloading a new container to bring down any other aircraft that poses an immediate threat.

Last moment self-defense

Aerial defense combines several overlapping systems, including long range missiles that can reach between twenty five and sixty miles; medium range ones, capable of striking targets within twelve miles, and short range ones that rarely reach more than six or seven miles. This last group includes portable weapons, capable of destroying any aircraft within four miles.

USED BY SAN MARCO

The infantry of the Italian Marines of the Reaggrupamente San Marco use their Stinger for self-protection when deployed in conflict zones; they have shown their complete confidence in their excellent capabilities for neutralizing all kinds of aerial threats.

The stinger replaces the Redeye

The United States Army had depended on large and sophisticated missile systems for its self-defense, weapons that guaranteed that not one enemy device could break through their protective shield. Self-defense, primarily, became the responsibility of the FIM-43 Redeye portable system that was first used in 1965 and was characterized by its small size.

Although large numbers entered service in no less than sixteen countries (the most extensive user was the United States with 13,000 of them) its limited aerodynamics, short range, speed of little more than Mach 1.6 and an explosive charge of just 2 kilograms meant that a new system needed developing with more powerful characteristics.

The Pomona Division of the General Dynamics firm was chosen for the task (they had already manufactured the previous one) and it was given instructions for development according to the requirements of the Army and the Marines. They asked for a system that could attack a target as it approached or left the position at which the launcher was situated; that could integrate an electronic IFF (identification Friend or Foe) system that could determine if an aircraft was hostile; that had a range of four miles; that was capable of following fighters and attack helicopters as they launched their own attacks, and that could, to a certain extent, distinguish between genuine targets and electronic decoys that were used to confuse the homing capabilities of an incoming missile.

In service in the early eighties

Combining the qualities of the previous model into a sufficiently light anti-air missile that could be transported without limitations by soldiers wherever they were sent was not an easy task. The FIM-82 was not ready until 1980, but proved to be a greatly improved weapon compared to its predecessor.

The first units of this anti-air model were made quite quickly and, in 1981, they had already been supplied to the American troops based in Germany, where they were used in all the exercises and maneuvers that were carried out. Their notable lightness and the positive reaction of all those who used them (which included the 82nd Air Transport Division at Fort Bragg, who received theirs in 1982) generated a great deal of interest among other Western armies. This model has a system that, at the time, was a yardstick when comparing all other similar models available on the market.

The Europeans formed a consortium to acquire it (which drew together all those countries interested in buying it) and make some of the parts under license, thus reducing the overall costs. These countries include Germany, Denmark, Italy, Holland, the United Kingdom (who used some of theirs as part of the Special Air Service (SAS) that was sent to neutralize the Argentinean air threat during the Falklánds Conflict), Turkey and France. France received one of the first batches

LIGHT MOUNT FOR TWO MISSILES

The Dual Mount Stinger (DMS) (produced by the Per Udsen Aircraft company in Denmark and by Raytheon in the United States) is a structure designed so that the operator can handle two light missiles more easily and effectively.

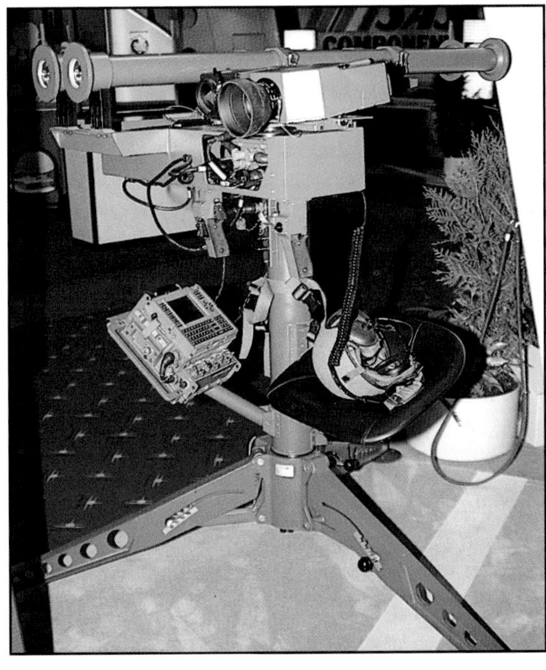

MOBILE AND EFFICIENT

The Avenger system has been configured for an electrically moved gun carriage that can handle two containers that hold a total of eight Stinger missiles in their firing position; this makes it possible to confront saturation attacks or advancing helicopter patrols.

SHORT RANGE ANTI-AIR

The armored LAV-AD is used by the Marines for short-range anti-air self-defense, using a 25-millimeter multi-barrel artillery mount and a quadruple gun carriage for Stinger missiles. These missiles can be substituted with Mistral if a particular client so desires.

because its own defense industry did not possess any such weapon. The process of marketing the weapon was slowed down by the restrictions of something classed as extremely sensitive because of its anti-air capabilities, and the subsequent fear that any guerilla or terrorist group would naturally be interested in acquiring the weapon for use on their violent anti-humanitarian missions However, the next buyers were Angola, Saudi Arabia, Chad, South Korea (who in 1997 ordered another 1,065 RMP type), Iran, Japan, Nicaragua, Pakistan, Thailand and Taiwan. Later Afghanistan joined the list (whose guerillas were armed with these missiles to face the Soviet aircraft that participated in the invasion of that country in the last decade) and Israel. This last country has regularly used its launchers in the ongoing conflict with its neighbors seeks to maintain the safety of the nation.

Easy to use and excellent features

This system, which is now made by Hughes Missile Systems and Raytheon, was developed to be easily used with the minimum of tuition. The basic criterion is that of fire and forget; the missile seeks the target on its own and once fired does not need any other kind of action. Although different training systems have been devised to teach its user how to operate it (such as those carried out by the marines in a large dome onto which a scenario is projected that includes several targets that must be fought by the person controlling the simulator situated in the central part of the area, from where he can seek and destroy the targets shown to him by the instructor) it is very simple to use and does not require much practice.

ASSURED MOBILITY

HUMMER vehicles mobilize the Avenger system, which has been positioned in the rear trailer and is made up of an operator sat in the middle of the gun carriage to direct it towards the area in which the target is located.

Real self-defense capabilities

The operator transports the fiberglass container (a little more than five feet long) that has the missile inside ready to fire. In the front left segment is the aiming element; on the right, a metal box that holds the IFF system and, beneath, the level that activates the launch; on the sides of the container are the protective plugs that protect the sensitive parts of the Stinger.

ALL WEATHER USE

The installation of systems like the AN/PAS-18 SNS (Stinger Night Sight) thermal camera made by Raytheon Systems Company means that this anti-air missile system can be used just as easily by day or by night. This module permits the operator to find the point in which the target is located.

Once these have been removed, the person who controls it must set it on his right shoulder and look through the visor until he sees the target aircraft; he locks onto it, makes sure that it is indeed the target and fires the weapon by making a slight upwards movement to compensate for the blast that is produced by the ignition of the rocket. The missile is autonomous up to the point of impact or the automatic explosion of its explosive warhead once it has passed 5,500 meters, which is considered the maximum distance it can be used for.

Once the launch has been made, another container can be installed and the homing unit and trigger can be used again. Because it uses reusable parts as many times as necessary, the costs of handling and use are reduced.

Features that correspond to its age

The rocket that provides the impulse uses solid fuel that is made up of an accelerator and a cruise engine that can reach a maximum speed of Mach 2.2 and height of 4,800 meters. The front warhead includes both the homing device and the charge. The former includes a passive infrared element in the A model and an infrared and ultraviolet one in the B model. The lethal element is made up of an HE (high Explosive) type warhead that has a 3-kg charge. It includes an impact fuse and has an effect radius of about five meters.

The body of the missile has a diameter of 7 centimeters and a wingspan of 9.14. Each one of the munitions weighs 10.1 kg, the size of which corresponds to the missile itself. The weight of the complete system is 16.1 kg. To improve its features, in 1982, the Stinger-POST (Positive Optical Seeker Technology) was introduced or FIM-92B that could make the target out against the background. In 1989, the RMP (Reprogrammable Micro-processor) came into service, also known as FIM-92C, which includes a reprogrammable microprocessor to face up to the threats of the future and is only given to those countries that the United States considers its strongest allies; the Block I or D that have been selected for Italy, and the Block II that will come into service early in the new century. This can install the Raytheon AN/PAS-18 system that is made up of a passive thermal camera called SNS (Stinger Night Sight) that can be used

LIGHT ANTI-AIR SYSTEM

The Germans have configured the ASRAD light anti-air system, made up of a quadruple pedestal for the Stinger and its observation systems installed on a small Wiesel 2 armored caterpillar that can be deployed where it is needed by heavy transport helicopter.

without limitations in even the worst climatic conditions. It only weighs 2.5 kg and launching is a simple question of connecting it to the upper section of the container.

Different deployment platforms

The qualities that have been demonstrated by this weapon (of which more than fifty thousand have now been manufactured) and the specific needs of some of the users have led to different configurations being made that satisfy the requirements of certain military groups or proposals that certain bodies have come up with. At the same time, it is worth remembering that it is now also possible to equip different helicopters with gun carriages for launching these same missiles in order to take out similar weapons being used on the ground.

One of the most outstanding terrestrial varieties is the Dual Mount Stinger (DMS) (offered by the Per Udsen Aircraft company in Denmark as well as Raytheon in the United States), which has a light structure that supports a seat for the operator, the platform that holds the electronic control unit and the day and night observation elements, as well as two complete missiles for confronting the most modern of threats (such as cruise missiles or remote controlled reconnaissance vehicles). The complete set weighs a total of 95 kg, including the weight of the terminal cambra that receives systems data from the command center.

Other terrestrial systems are the ASRAD (Short Range Air Defense) made by the German STN ATLAS Elektronik company (which includes a four-missile pedestal installed on the chassis of an MaK Wiesel 2 caterpillar vehicle), of which 50 platforms have already been ordered by the German army and should all be ready by the year 2003; the Israeli Machbet, which mounts a quadruple container on the gun carriage of a 20 mm multi-barrel Vulcan M183 gun placed on an

HIGHLY DEVELOPED GERMAN ANTI-AIR

The Daimler-Benz Aerospace company designed this light two-legged support that lets the operators control two Stinger and the target sight unit. It is easily deployed by vehicles in any zone that could be threatened by aerial threats.

TECHNICAL CHARACTERISTICS FIM-92B ANTI-AIR MISSILE SYSTEM

COST IN DOLLARS:	300,000
DIMENSIONS OF THE MISSILE:	
Length	1.52 m
Diameter	0.07 m
Wingspan	0.091 m
WEIGHTS:	
Of the missile	10.1 kg
Of the firing unit	6 kg
Of the explosive charge	3 kg
PROPULSION:	
Two stage rocket engine with solid fuel	

FEATURES:	
Maximum range	5,500 m
Minimum range	200 m
Maximum operational height	4,800 m
Minimum usable height	30 m
Maximum speed	Mach 2.2
Probability of impact	91%
Nocturnal use	Yes, with thermal camera
Shots per minute	Needs 15 seconds for each reload plus the time for realigning and firing
CREW:	One shooter and one assistant

M113 caterpillar vehicle, and the Turkish Atilgan and Zipkin, developed by Aselsan Microwave and Technologies System Division; the latter includes an eight-missile launcher on an M113 or one with four missiles on a Land Rover Defender respectively.

Even more advanced, owing to further design improvements, are the United States Avenger, M6 Linebacker and LAV-AD systems. The former is a light 4x4 HMMWM vehicle that moves a 12.70 x 99-mm electrically powered pedestal; Boeing Defense has made more than a thousand for the Army and the Marines. The second is a development of the Bradley infantry combat vehicle on which the TOW anti-tank missile launcher has been replaced by four Stinger; since 1997, the Army has received 267 units. The third is a model that was specially designed for the Marines, who ordered 17 of these systems that are formed by an 8x8 LAV light armored vehicle that transports a turret with four missiles and a 25 mm GAU-12/5 multi-barrel gun.

MORE THAN A THOUSAND IN SERVICE
The United States Army, Marines and National Guard own more than a thousand Avenger systems that make up the most up-to-date section of United States anti-air technology.

USED BY THE MARINES
The Marines have every faith in the Stinger for anti-air self-defense. Operators work in pairs, with one operating the launcher and the other studying the sky with binoculars in search of targets. The latter also operates the radio to receive data about possible threats.

The French Armed Forces' need for a light missile system that would provide them with the necessary defense in any circumstances, and the lack of any similar weapon in the extensive French defense product catalog, led the technicians of Matra to begin developing a new missile that came to be called SATCP, corresponding to the French Sol-Air-Très Courte Portée (Short Range Anti-Air System).

Its design was based on the "fire and forget" concept, which meant fitting it with an autonomous infrared type self-guidance system, with the necessary miniaturization to contain all the equipment in a very small fuselage and with provision for keeping it on a gun carriage that would provide the maximum comfort.

The Mistral is born

The first generation of portable light anti-air missiles did not affect the French designers' program, and they continued working on their medium and long range systems. Coinciding with the arrival of the first second-generation units (with far superior features to their predecessors), the French military realized that they needed a similar type of system.

HIGHLY MOBILE

In a special backpack (conventionally folded and set on the operator's back), the launcher of the Mistral anti-air system. The box that the soldier is carrying in his hand contains the cooling elements that are needed by the homing devices.

SEEKING NEW MARKETS

The Guardian is a Matra Bae Dynamics proposal that is made up of a highly mobile HMMWW vehicle and a Boeing gun carriage for launching six Mistral missiles. Its system configuration is very similar to that of the United States Avenger.

Ten years to prepare it

The representatives of the staff of the French Armed Forces and members of the Délégation Générale pour l'Armement (DGA) formalized a study commission in 1977 to define the characteristics of the weapon needed to counter the aerial threats situated at short distances. After two years of work, the list of requisites that such an anti-air missile system needed to comply with was complete and called SATCP. This had to satisfy the requirements of the Army, Marines and Air Force, and five companies offered to develop it.

Having examined the five possible programs, in December 1980, a contract was signed with the Matra company for the development of the system within six years and to work simultaneously on the design of the different platform options onto which it could be installed. The first terrestrial and naval launchers were ready within the proposed time limit, and were validated by an extensive test program, which included launching a considerable number of missiles. The first units of this series were introduced to the French forces in 1990.

The most outstanding characteristics of the missile include the fact that it can reach a maximum speed of 835 meters a second. It includes a 3 kilogram warhead that projects 2,500 tungsten bullets either when the proximity fuse is laser activated or by impact. It is capable of striking targets at distances of up to four miles and it takes just twenty seconds to reload.

IMMEDIATELY READY

It is usual for each Mistral anti-air system to be operated by three men who are in charge of transporting it, setting it up in the most suitable position and controlling its anti-air function. One of them provides rifle cover for the other two as they handle the launcher.

Internationally successful

From the more than 500 real-life missile shots against all types of target (which include shots from Alouette II helicopters flying at top speed towards Chukar type targets) it has been calculated that it has a strike rate of 93 %, which could be increased to 94 % if only those shots made since 1995 are taken into account.

These remarkable aerial target interception capabilities, derived from their high speed and powerful explosive charge, have led to more than 12,000 units being purchased, now sold by the French-British Matra Bae Dynamics consortium. Among the buyers of the missile are 33 different armed forces in 21 countries, made up of 8 in Europe, 7 in the Asian Pacific, region 3 in the Middle East and 3 in Latin America; including Brazil, Chile, Cyprus, South Korea, Spain, Finland, Indonesia, Qatar, Norway, New Zealand, Singapore and Thailand.

Despite the secrecy that surrounds the sales of this system and the fact that the

manufacturer does not offer exact details about its export, the Consejo de Ministros of the Spanish Government authorized, on the 13th of December, 1991, the acquisition of 200 launchers and 800 missiles at a cost of 100,000 dollars. A sizable part of the components of these was made by Spanish companies, which means that there is a 90 % return on invested capital. Twelve of these systems went to the Brigada de Infanteria de Marina and the rest to the Ground Forces. Further launchers are expected soon, which will incorporate a thermal camera for use in all weather conditions.

CAPABLE OF DEFENDING ITSELF

The Mistral system gives the infantry units their own short-range anti-air self-defense. Two men are needed to operate it, one moving the launcher and activating the missile and the other in charge of communications that warn of the possible approach of targets.

Multiple launch platforms

Among the requirements that were considered by the military at the beginning of the development process was the fact that it should be possible to use the system from terrestrial carriages, on vehicle assemblies, on naval mounts, and as a form of self-defense to use on board helicopters. Therefore, a wide range of different mounts has been designed from which it can be fired.

Adapted to every need

The manufacturer's proposals include a portable launcher, used by the infantry troops to defeat targets that approach by converging routes or at any angle (except vertical, because the blast would injure the user). This is made up of the launcher, pointing system, cooling element and the missile containers.

The Atlas has greater capacity and weight, moves two missiles and can be placed either on the ground or on vehicles. The Aspic is similar to the previous model, and has a built-in visor that only has to be turned towards the target so that the launcher (positioned a few meters away) can aim automatically towards the target. The Santal is highly mobile, it is autonomous with respect to the detection of aircraft thanks to its Doppler pulse radar, and incorporates six missiles. The system can be installed on different types of wheeled or tracked armored vehicles. The Guardian, an evolution of the Avenger, is also mobile.

The co-ordination of the Mistral launchers can be done from the Mistral Coordination Post, which incorporates exploratory radar and display screens, and at the same time the Atam gun carriage is proposed for the self-defense of helicopters. The Atam is already used on Gazelles and will soon appear on the Tigre too. Simbad light twin launchers can be

ATLAS SYSTEM

The Spanish Ejército del Aire has chosen the ATLAS system as a means of self-defense that can be set up easily wherever it is needed. This way, just one operator can handle the system, which is made up of two ready use missiles.

MORE MOBILE AND CAPABLE

The Santal is an anti-air system installed on a Sagaie armored vehicle with one turret, where the search radar that detects for targets is situated, and two triple carriages for short-range Mistral missiles that have the task of destroying them.

used on ships, as can the Sadral, which is sextuple and used on a stabilized launcher, and the Sigma, which uses three missiles on a 25 or 30 millimeter gun.

Spanish ATLAS

The first arrived at the Zaragoza base on board a C-130 Hércules on 24 April 1996. The system deployed by the Escuadrilla de Apoyo al Despliegue Aéreo (Aerial Deployment Support Squadron) is made up of four twin launchers updated to combat fighter-bombers or helicopters operating at high speed a low altitude. They can fight two targets at the same time because the missiles follow their targets independently.

These systems can be controlled from fixed positions on the ground or from the trailer of Nissan Patrol ML-6 vehicles. Its elements are a regulated base that weighs 20 kg and integrates five jacks that adapt it to the terrain; a 30 kg pedestal that supports the operator's seat and the battery; an upper element that includes the two missile containers; the triple magnification fire command visor; the cooling modules for the homing devices and the IFF (Identified Friend or Foe) interrogator.

ANTI-SATURATION CARRIAGE

The ATLAS (Affût Terrestre Léger Anti-Saturation) is a twin Mistral launcher that can be used either from the ground or on different vehicles. It has two munitions for immediate use in the gun carriage itself.

TECHNICAL CHARACTERISTICS LIGHT MISTRAL MISSILE

COST IN DOLLARS:	**237,000**
DIMENSIONS OF THE MISSILE:	
Length	1.86 m
Diameter	9.25 cm
WEIGHTS:	
Of the missile	19.5 kg
Of the warhead	3 kg
PROPULSION:	
	Solid fuel engine that ends combustion in 14 seconds

FEATURES:	
Maximum range	6 kg
Maximum height	3 km
Maximum speed	Mach 2,5
Daytime use	Yes
Nocturnal use	Yes, with thermal camera
Probability of strike	93 %
Velocity of lens	From stationary to Mach 1.2
Firings per minute	More than two
CREW:	**2 or 3 men**

MUNITIONS CONTAINER

Mistral missiles come from the factory in a sealed fiberglass container that protects them until they are ready for launching. In the upper part, they have a protective module that covers the homing device, and in the lower section, this covers the gas exhaust of the propeller.

CONTROL HELMET

The systems operator wears a protective helmet on his head with an integrated earphone through which he hears the signals that relate to the homing device inside the weapon or the orders that he receives from the command control center.

COMFORTABLE WORK

The operator of a light Mistral system carries out his work from a sensibly comfortable position, because the search and fire process is done from a seat attached to the right hand side of the launcher. If there were not this facility, the operators would soon be exhausted by the wait.

GUIDANCE AND CONTROL UNIT

This sight and related panel allow the shooter to center the target, and therefore he can work out the best moment to fire the missile at it. Learning to use this autonomous anti-air system does not take very long at all.

HOMING DEVICE

At the front of the missile is a homing device is the form of a prism that includes the capturing elements that pick out the thermal signature of the aircraft it seeks. This one is part of a training missile.

COOLING ELEMENT

When the homing device is activated, so is a battery that acts as its cooling system during the lock-on period before firing. It only lasts for forty-five seconds, but also supplies energy to the system.

360° ROTATION

The light Mistral's gun carriage was designed so that the operator can turn it quickly trough 360° without becoming tired, so that he can cover any aerial threat to his position.

LIGHT SUPPORT

A vertical column and three supporting elements on the ground are enough to stabilize the Mistral launcher, which assists the operator in his work when aiming and firing the missiles.

The professionalization of the British Armed Forces brought about a reduction of numbers and the adoption of the most modern arms systems to keep up its military potential both in the British Isles and in the Overseas Territories they owned, or still own, and where they have economic and sovereign interests.

This policy (maintained by different governments since the Second World War) has led to the development and manufacture of all kinds of equipment, and its powerful industries cover the majority of combat needs. The United Kingdom was one of the first countries to notice the danger of different types of aircraft to terrestrial units. Therefore, development began on an advanced group of SHORAD (Short Range Anti-aircraft Defense) systems based on portable short-range missile launchers.

Development begins at the end of the seventies

The British Army wished to supply their troops with a light missile system that could destroy attacking aircraft before they had the chance to launch their weapons. Therefore, they requested various companies to present proposals for the development of the weapon. Eventually, the design and development work was put into the charge of the missile systems division of the Short Brothers Company in Belfast.

VERY EFFICIENT WARHEAD

The Javelin and some of the models that have followed it include a highly explosive warhead that can bring about the total destruction of light helicopters and cause serious damage to larger aircraft.

WITH THE ROYAL MARINES

The British Marines, working with the Dutch and an Amphibious Brigade together, deploy their Javelin triple launchers to protect their beachhead movements from any enemy reaction from the skies.

Deployment of the Blowpipe begins

After several years of investigation and development activity, the anti-air system was completed. It was made up of a container tube that held the missile and the independent aiming unit. In 1975, the first units of the Blowpipe, as it was named, entered service with untis who evaluated the new anti-air weapon favorably. It was subsequently given positive reviews in countries such as Afghanistan, Argentina, Canada, Chile, Ecuador, Malawi, Nicaragua, Nigeria, Oman, Pakistan, Portugal, Qatar and Thailand, who received several batches of this weapon until there were some 21,000 missiles in circulation.

Carried on soldiers' shoulders or set on the quadruple turrets of a Spartan caterpillar vehicle (mounting it on a submarine was even considered), these missiles, capable of reaching speeds of Mach 1.5 and ranges of 22 miles, demonstrated outstanding characteristics for their time. Of these features, it is worth noting that it was not guided by heat signals from the target and that its course could be controlled by the operator himself to increase the chances of the target being struck. This is made possible by the aiming unit which incorporates a radio transmitter that operates on several frequencies selected by the person who controls it, a monocular sight and a control device. These devices work so that the missile (after being fired from the launch container in which it was sealed) is automatically directed towards the center of the operator's field of view, who guides it towards the target by means of a radio control. The explosion of the 2.2-kilogram warhead is produced either on impact or when the proximity fuse is activated.

The experience of combat

During the Falklands Conflict in 1982, the British deployed several anti-air systems on the islands, including Blowpipe surface-air missiles that, despite weighing 21.9 kg, were transported by the anti-air expeditions that advanced on foot with the rest of the troops. This relative lightness, which meant it was possible to deploy some of them on the decks of ships, made it much easier to move forwards against the threats of Argentinean air attacks. According to the British Ministry of Defence, nine, and maybe even more, were brought down.

TWO OPERATORS

To employ the firing capacity of these short-range missile launchers developed by Shorts Missile Systems, the soldier who manages the system is joined by a colleague who studies the horizon with binoculars to find out the area and height of enemy craft that must be eliminated.

Argentina also used them and it seems likely that one of them was responsible for bringing down a Harrier attack airplane whose pilot ejected and was captured.

However, the weapons were still limited in capabilities when it came to dealing with aircraft flying at high speeds, and so improvement programs started. These resulted in a new model that received the name of Javelin. This, produced since 1988, incorporates a more

ASPIC SYSTEM

The ASPIC system includes an eight-container carriage for Starburst missiles and a guidance unit in the center. They can be launched at the target in sequence so as to increase the chances of destroying it.

advanced guidance system called SACLOS (Semi-Automatic Command to Line-Of-Sight), made up of a statimetric visor with six magnifications and a television camera that centers the target and transmits orders that the missile uses for semi-automatic guidance towards the line of sight. At the same time, other features have been improved; the propellant now enables missiles to reach targets at distances up to 5,500 meters and at altitudes up to 3,000 m, and the explosive warhead contains 2.74 kg of explosives; this increases its overall weight by 2.4 kg compared to the Blowpipe's 21.9. These new characteristics, and aggressive marketing on the part of the British defense industry, have led to the sales of 16,000 units of the light variety and with a triple carriage to Great Britain and three others. Of these three, only the South Korean purchase was announced to the public.

GOOD CAPABILITY OF RESPONSE

Javelin triple missile launchers enable an excellent capacity for response to saturation attacks, because they have missiles designed for immediate use. This avoids the reloading operations of simpler launchers.

TECHNICAL CHARACTERISTICS STARSTREAK ANTI-AIR MISSILE

COST IN DOLLARS:	250,000
DIMENSIONS OF THE MISSILE:	
Length	1.397 m
Diameter	127 mm
SPAN OF THE LAUNCHER:	274 mm
WEIGHTS:	
Of the missile	13 kg
Of the guidance unit	11.9 kg
WARHEAD:	
Three highly charged kinetic darts that contain highly explosive charges that are activated by an impact fuse	

PROPULSION:	
Solid fuel rocket engine made up of the accelerator and the cruise propellant	
FEATURES:	
Maximum range	4.5 miles
Maximum operational height	2.5 miles
Daytime use	Yes
Nighttime use	Yes, with thermal camera
Probability of impact	Over 90 %
Maximum speed	Mach 4
Shots per minute	One
CREW:	1 man

Improving the systems capacity

Shorts Brothers (who had been given the job of perfecting the Javelin at their Belfast factory in Northern Ireland) initiated several study programs to increase their range of short-range anti-air missiles, introducing new variants with greater capacity and characteristics. This way, it would be possible to confront the latest generation and future helicopters and airplanes.

The Starburst is introduced

From 1990, the Javelin has included a laser guidance system which is immune to interference and is known as Starburst. It was used during the Gulf War, and more than 10,000 units have been produced for use by the British Armed Forces and also those of Canada, Kuwait, Malaysia and Qatar. This missile, whose manufacturer claims is immune to every known type of interference, is transported in a fast-employment unit that is made up of a launch tube that is rested on the soldier's shoulder and a system that is used for guidance. This weapon is able to reach a speed of Mach 2 and a distance of four miles. As it travels towards its target it is sent on course by the four guidance fins on the tall which work in conjunction with a further four smaller fins at the front or the missile, just behind the warhead.

DAY OR NIGHT USE

A group of British marines is preparing a Javelin /Starburst system triple launcher, taking advantage of night conditions to avoid detection and, when day breaks, they will be able to fight off air attacks. When being deployed, the missiles are kept in plastic containers that protect them from accidental knocks.

EASILY TRANSPORTED

The light tripod that supports the Starstreak anti-air defense system can be easily transported to the zone of deployment, adding the containers of missiles and the guidance unit to it beforehand.

To obtain a more capable weapon than the previous one, a contract was signed with Shorts on December 15th 1986, for which £225 million were invested in designing the HVM (High Velocity Missile), which is commercially known by the name of Starstreak. This can be situated on triple vehicle carriages such as the Aspic or on more sophisticated ones that are moved on platforms formed by the Alvis Stormer armored vehicle. It is so advanced that in April 1989, a British Citizen from Ulster, who it seems was working with the South Africans, was arrested for attempting to get hold of information relating to its characteristics and performance capabilities.

Up until now, the self-propelled variant has been used by the British Royal Artillery, which includes eight HVM in firing position, a passive ADAD warning device made by Polkington Optronics and the operator's pointing visor that uses laser guidance to reach the target.

The Starstreak has been evaluated by the United States Army and fired from an AH-64 Apache helicopter at its Yuma site to check out its characteristics as an aerial self-defense weapon. The company Lockheed Martin promotes its installation on helicopters both in the United States and in the United Kingdom.

EXPERIENCED IN COMBAT

British Javelins have inherited the combat experiences of the Blowpipe during the Falklands Conflict. It has been substantially improved with regards to ease of use and operative potential.

Future programs

At the beginning of 1999, the British Ministry of Defence signed a contract with Short Missile Systems (SMS) for the development and provision of thermal vision equipment for the self-propelled variant of Starstreak. This equipment, which depends on the evaluation of five different companies' proposals, will enter service in 2003, and SMS will be in charge of carrying out integration and verification tests on its actual characteristics.

The HVM Stormer should be able to detect, recognize, identify and pursue its targets by night and in adverse weather conditions, thus overcoming the deficiencies of the current model, which can only be used by day and in clear weather. All of these will be unified in one regiment, which will have 36 systems in each of its three light missile anti-air batteries. Current expectations predict that the 135 self-driven mounts used by the British Army will be upgraded to this new standard, and maybe they will also receive a thermal camera and an IFF (Identified Friend or Foe) system in the Starstreak triple launchers. This would increase their current potential so as to bring them in line with the needs of the new millenium.

A variant of the Stormer has been proposed for export, which combines eight Starstreak launchers with a thermal vision visor and one for automatic pursuit. This allows for the configuration of an anti-air missile system with a range of between four and five miles that could replace other similar systems developed as mobile air-defense mediums.

ALL-WEATHER USE

Rain, which can be seen on this British Starstreak anti-air missile system, does not hinder the operability of the weapon as long as the aircraft to be attacked is still within sight.

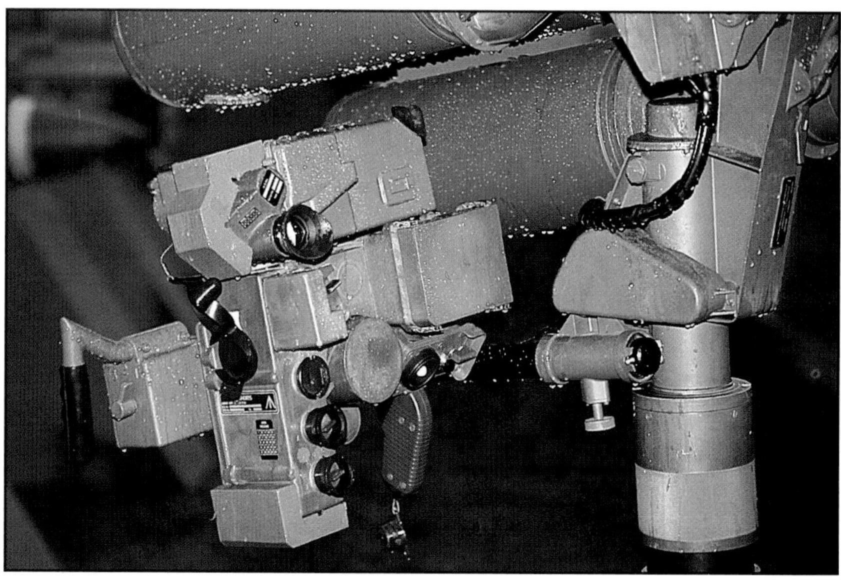

The need for a missile system against low-flying aircraft that could replace 40/70 mm Bofors in the British Army led to the design and adoption of the Rapier missile. This has been widely exported and even used in combat during the war between Iran and Iraq and during the Falklands Campaign.

This weapon, which is available in towed or self-propelled varieties, has been optimized to deal with low-flying aircraft that come within range, even in air that is saturated by electronic counter-measures.

Thirty years in service

The first scale models of what would be a new anti-air system based on the use of missiles took its first form at the design studios of the British Aircraft Corporation in 1961. In 1964, the British Ministry of Defence published their specifications for the required model. To reduce the weight and adapt it to these requirements (which needed a reduced volume to favor its transport) an manually guided optic system was chosen that guaranteed, according to the manufacturer, that the missiles would strike the target directly, for which reason the proximity fuse was elimi-nated and the size of the explosive charge reduced.

COMPACT MODULE

The towed module weighs 2,400 kilograms, is 4.1 meters long, 2.2 wide and 2.6 high. It includes a launcher and eight missiles and an electro-optic surveillance and pursuit system.

TRIED IN COMBAT

The Rapier anti-air missile system has been used in combat both in Iran and in the Falklands Conflict, where it has proved most effective, particularly considering the fact that it is much cheaper than many similar systems.

Acquisition is agreed

The excellent features of the prototypes that made direct strikes on a wide variety of aircraft (including towed Rushden decoys of only 19 cm diameter) led to them being adopted by the British Army, who received the first towed models in 1970.

The system (which could be transported on two light vehicles) was made up of a launcher that included four side carriages for that number of missiles; in the upper section, search radar inside a dome that crowned the mount; at the front, the radar which followed the missile and the control set with a unit for selecting dead angles for the launcher and a

Barr & Stroud optic pursuer that was used for aiming the missiles.

Its remarkable features, which had the systems and capacity for flying at Mach 2, enabled it to reach the fighter-bombers in service at that time without any problems. 13 countries purchased the new weapon. These included Abu Dhabi, Australia, Brunei, Iran, Oman, Qatar, Singapore, Switzerland, Turkey, Zambia and the United States, who bought some batteries to defend its bases in the United Kingdom.

More developed versions

The sales success, which encompassed some 20,000 missiles and 400 firing units, inspired the development of an all-weather variant that was to include radar for use at night and in adverse weather conditions.

In 1975, the Marconi company put the finishes touches to the DN-191 Blindfire system that consisted of towed radar that combined an antenna and a television camera that assisted the guidance of the weapon towards the target. This action was carried out automatically, enabling the missile to strike the target accurately. The Rapier all-weather unit requires three light vehicles to be moved.

An order that arrived from Iran (governed then by the Shah of Persia) helped with the development of a self-propelled variant to advance with and protect armed units. It was transported by an M-548 armored caterpillar vehicle. After the fall of the Shah and owing to the enormous amount of money that had been invested, the British Army ordered 64 units which were received from 1983. The outstanding features of these units were their octuple launchers and the optic homing device situated on top of the cabin. This device was improved with time through the addition of a thermal search device.

THE JERNAS MODEL

The need to confront the air threats of the new century led the technicians at Matra Bae Dynamics to configure the Jernas anti-air system, basing its capacity to destroy any kind of air threats on its high level of automation, which means it can act as an immediate response.

The development of the system continued, and taking the Rapier 90 as a precedent, Rapier Laserfire was created, which included a single launch mount, search radar, a two-man cabin and a laser guidance module.

Rapier Darkfire was next to appear, used by the British Army since 1988. At the same time, more advanced missiles were developed, including the B1, the B1X (which was a development of the first but now incorporating digital technology), the Mk 2, the Mk 2A (with a redesigned explosive head) and the Mk 2B, which included a fragmentation charge.

At the beginning of 1999 an 81 million dollar contract was announced between the Swiss government and Matra BAe Dynamics for the production phase of the Swiss MidLife Improvement (SWIMLI) program for the 60 Rapier that had been used since 1986 by the Swiss Air Force. This was complemented

GUIDANCE RADAR

The Marconi DN-181 Blindfire is towed radar that combines an antenna and an optronic system that guides the missile towards its target, thanks to the continuous pursuit that allows their routes to cross over.

TECHNICAL CHARACTERISTICS RAPIER FSC ANTI-AIR MISSILE

COST IN DOLLARS:	Unknown
DIMENSIONS OF THE MISSILE:	
Length	2.24 m
Length of firing position	4.1 m
Height of firing position	2.6 m
WEIGHTS:	
Of the missile	43 kg
Of the firing position	2,400 kg
WARHEAD	
Fragmentation type with a laser activated multi-mode proximity fuse	

FEATURES:	
Maximum range	5 miles
Maximum operational range	3 miles
Maximum speed	Mach 2.5
Maneuverability	30 g's
Daytime use	Yes
Nocturnal use	Yes
Shots a minute	7
Reloading time	2 minutes
Probability of impact	+90 %
CREW:	6 men

by the modernization of the surveillance, launch and guidance radar that was contracted to the British company in May 1995. The SWIMLI is similar to the export version of the BIX Rapier, but not does not include the latest British-designed Rapier Mk 2 missile, which over the next few years will replace the self-propelled variant that is used in Starstreak.

The latest in air defense

The merger of British Aerospace with the French company Matra has led to the reorganization of the products that they both offer, making it more rational and competitive, and more receptive to the needs of the market. As a result, this system has evolved into the concept of Jernas, which is based on the use of the anti-air Rapier FSC missile used by two batteries of the British Army's Royal Artillery and by three US Air Force squadrons.

Concept of traditional deployment

The Jernas was created to confront the latest and more varied types of air threat, which include cruise missiles, remote control reconnaissance vehicles and very low flying helicopters in any kind of weather conditions and despite any electronic counter measures being used. This is made possible by the use of an automatic system that has been tested according to the strictest British government specifications. It is characterized by its excellent mobility whether being towed by one of

GUIDANCE OPTRONICS

Jernas launchers integrate a large sphere that contains all of the autonomous electro-optic surveillance system that operates passively. It serves to improve the accuracy of the pursuit of the missile (right photograph).

IMPROVING ITS CAPACITY

The latest versions of the Rapier include a surveillance radar module that weighs 2,200 kilograms and is fixed to the tow vehicle to make it easier to move. It includes an antenna that emits Doppler pulses and turns at any selected speed between 30 and 60 r.p.m (right photograph).

many types of vehicle, or being transported by one of an equally large number of aircraft.

Its superiority is based around the rapid detection and destruction of an enemy threat; it can pursue its target very accurately; in operative mode it is totally automatic; it is largely immune to most known countermeasures; it includes several sensors that work simultaneously, and it has both active and passive operation modes. All these characteristics help it to survive anti-radar missile attacks, which are designed so that its crew is separated from the system by the protection of their armored vehicle.

The Tactical Operations Cabin (TOC) includes air conditioning, is easy to control, has automatic testing systems and can be used 24 hours a day.

All this has a positive effect on the efficiency of a system designed to work in temperatures ranging from –36° to +50°, and even in extremely muddy or sandy conditions.

The modular structure of the system increases its chances of survival and means it can be easily moved to any new location. This system can cover a 360° area in which it can detect threats at up to 10 miles away and eliminate any that come within a perimeter of 5, be it day or night and regardless of weather conditions.

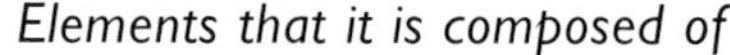

PROTECTED FROM THE ELEMENTS

The surveillance radar associated with the Jernas version of the Rapier is set within a plastic dome where it is protected from the weather and which also manages to keep it hidden from enemy observation elements.

MODULAR CONCEPTION

The fact that the Rapier has been designed as a set of towed parts that include the subsystems of this anti-air system make it easier to move by land or by air. At the same time, this makes it less vulnerable, because the different components can be kept apart.

Elements that it is composed of

In order to achieve these excellent capabilities a system has been designed made up of surveillance radar, guidance radar, the launcher, the electro-optic guidance and surveillance system and the missiles.

The former is what provides the system with details of the targets in three dimensions, amongst which are included airplanes with low radar signatures, to a number of up to six launchers connected to the system by a fiber-optic link.

With a weight of 2,200 kilograms, it works on band J, includes an automatic frequency jumping mode and is able to locate up to 75 targets a second provided they are within a radius of 10 miles and a height of 3 (which is the area covered by the rotation of its antenna that moves at an alterable speed of between 30 and 60 r.p.m). It is able to detect an intruder that enters its area in less than 1.5 seconds, which it will interrogate with an automatic IFF Mk10 or 12 to check whether it is friendly or an enemy.

The guidance radar (monopulse type, and operating on frequency band F) can pursue the target in absolutely any kind of weather, acting automatically so as to strike a target that comes anywhere within its 10 mile range. It weighs 2,600 kilograms and is 2.64 meters high.

As for the launcher, it combines eight missiles that can be fired extremely quickly (the reaction time in automatic mode is of five

RAPIDLY DEPLOYABLE

The small size of the octuple launchers of the Rapier anti-air system make it easy to transport it to the area it needs to defend, which might be an airport or some kind of important terrestrial installation.

seconds for the first target and a further three for the second) so that up to seven can be launched in just a minute.

The same launcher incorporates a passive infrared surveillance element, giving it the exceptional possibility of operating autonomously without needing the help of the aforementioned elements.

Finally, it is worth mentioning that the missile includes the very latest digital technology so it can even be used against saturation attacks. It can reach further than five miles, its maneuvering capability is 30 g's and it uses ACLOS guidance, which takes the pointing line indicated by the radar sensor as its main point of reference, and uses the optronic one as the second.

HEAVY FIRE CAPABILITIES

The most recent Rapier launchers integrate four missiles on either side that are ready to be fired as soon as the order is given. To confront saturation attacks, seven of these missiles can be fired in one minute.

Originating from a collaboration between France and Germany, this anti-air missile system has enjoyed a healthy reputation after being used during the Falklands War to protect the area of Puerto Argentino. The Marine infantry are convinced that, with just eight missiles, four definite strikes, and probably one other, took place, as well as taking out a free falling bomb.

Whether or not this Argentinean data is correct (the makers of the system have extensively used the statistics in their marketing), there is no doubt that the Roland is an extremely robust and capable element. It has been in constant development since the start, ending up as an anti-air defense that pilots fear heavily. This says a lot for its features and ability to generate fire in even the most adverse of weather conditions.

SPANISH ARMY

No 81 Anti-Air Artillery Regiment of the Spanish Marina at Valencia, assigned to the Maneuver Force, includes 16 self-driven Roland Group units on the modified chassis of an AMX-30EM1 tank.

Joint development

A political arms agreement, signed in 1962 by representatives from France and Germany, set off this industrial collaboration for the development of different missile systems. These included one for a short-range anti-air system. Inspired by the former, two years later work began on developing an optically guided system that would be able to operate whatever the atmospheric conditions. This work was put under the charge of the German Messerschmitt-Bölkow-Blohm company and the French Aerospatiale.

HIGHLY MOBILE SHELTER

This configuration of the Roland, exported to the Middle East with a different color on the outside, is designed to be able to move independently to the action zone. This particular model is noted for its elastic group and the air conditioning unit situated at the rear.

The model to follow is decided

Coinciding with the development of the prototype, which was completed in 1969, a new generation of airplanes and helicopters was starting to appear, designed to attack either by day or by night. The collaboration between both companies was consolidated by the formation of a joint company EUROMISSILE, who took on the challenge of perfecting a version that could not only operate in all weather conditions, but could also satisfy other tactical requirements.

After the evaluation tests on the Roland I (as the fair weather variant was known) and on Roland 2 (the name of the all weather variety that was completed in 1971 with guidance radar being added to the original model), production of the system could begin. This consisted of 340 units of the latest version and 12,000 missiles for Germany (which were to be used on the chassis of the Marder infantry combat vehicle), as well as 214 units and 10,800 missiles for the French Armée de

Terre. The French Army used theirs on the modified chassis of an AMX-30 medium tank and incorporated two thirds of them as the first model and the remainder as the second.

In 1973, the United States Army hired four Roland 2 for evaluation as part of the SHORADS (Short-Range Air Defense System) program. It was indeed chosen, and EUROMISSILE subsequently received 265 million US dollars for providing production rights to the American Boeing and Hughes Aircraft companies.

The presentations begin

In 1977 the first Roland 1 were unveiled, and in 1981, the French and German Roland 2. The success achieved during evaluation (which revealed a system with the potential to destroy 80 % of its targets) pushed its sales. Among these were included countries like Argentina, who received a version in a container that was used in the 1982 Falklands Conflict; Brazil; Spain, who have 18 launchers positioned on AMX-30EM1; Iraq; Nigeria; Qatar and Venezuela. The total number of units produced reached 680 launchers and 26,000 missiles.

As for the US Army, they began making their own in 1979 and manufactured a total of 27 units and 595 missiles until, in 1985, Congress ordered production to cease. They were assigned to the National Guard in New

GERMAN ROLAND

The German Armed Forces use Fla Rak Rad systems for the defense of vital objects, such as air bases. This Roland carriage is noted for being positioned in a container transported by a MAN 8x8 truck.

CATERPILLAR CHASSIS

Both the AMX-30 tank and the armored combat vehicle of the Marder Infantry have been configured to mobilize Roland anti-air missile systems, a design that allows them to advance along with the armored and mechanized units that should provide protection from different types of air attacks.

Mexico, who started off using them from a container set onto a 6x6 M812A1 truck, and later from the chassis of a modified self-propelled M109 howitzer. The US Air Force, on the other hand, acquired the Fla Rak Rad Roland 25. These were also acquired by the German Air Force and Marines, who used a sheltered version for defending their respective bases.

In 1985, EUROMISSILE announced an improvement program that would lead to the Roland 3, incorporating a faster missile, with longer range and greater capacity with respect to its explosive charge. From 1992, a Roland M3S variant has been in development, using new supersonic missiles, an IR/TV pursuit system, multi-purpose screens, and 3D-surveillance and pursuit radar. Although there is no information of any sales of the former, what is known is that since the early nineteen nineties, the French and German firing points have been undergoing modernization that will keep them operative until 2010. Moreover, some of the original systems have been installed as sheltered versions that can be transported on trucks. This is the case of the 20 French ACMAT semitrailers and 10 cabins on German 6x6 MAN trucks, launchers destined to comply with rapid deployment force requirements. As for the Armée de Terre, in 1998 they signed a contract with EURO MISSILE for the updating of 72 units into the Carol (Cabines Roland Tractées par un VLRA) configuration that will be assigned to the FAR and presented between 2001 and 2005. The revaluation, which will convert them into the M3Vs variant, includes the integration of the VT1 missile, with anti-missile capabilities and a range of 7.5 miles (it is the weapon that the latest generation of Crotale use); it has a Glaive optronic visor and the capacity to be used as an integral part of the Martha command system. It could stay in service until the year 2015.

SELF-PROPELLED VARIANT

The German army uses about two hundred Roland 2 systems, which it moves upon the chassis of a Marder Infantry caterpillar combat vehicle that has been modified for the new anti-air missile. A small batch has been exported to Brazil.

VT1 MISSILE

The weapon fitted to a Roland sheltered launch unit is a new missile that is noted for being faster, more agile, further reaching and having a better combat charge than the original weapon. These characteristics have persuaded the French to buy it with the aim to maintain their system active until 2015.

Firing capacity

The characteristics of this compact and integrated anti-air system are defined by its very conception, designed to fit the elements for guidance, detection and launching onto one carriage. Inside the shelter or vehicle are all the consoles necessary for its control.

Basic elements

The three angular sections of the Roland are the missile, the acquisition unit and the firing unit. The former consists of a cylindrical body with a conical warhead to which four angled fins have been attached, which mark the way it turns on its axis during flight so as to stabilize it without the need of other elements. It is worth pointing out that its rocket acceleration engine first drives it at a speed of 570 meters a second, and then the cruise engine keeps its speed up until its multiple effect breaking charge (made up of different radial effect hollow charges) explodes. The guidance system incorporates an automatic pilot that receives signals from the tele-controlled receiver to follow the assigned route. The firing unit makes sure that this happens. This includes the Thompson-CSF surveillance and mono-impulse pointing radar that follows the target and the radio frequency transmitter fitted at the rear of the missile.

The former unit also includes an optic pursuit device that facilitates the guidance of the missiles by day thanks to its LOS (Line Of Sight) channel. This moves the path of the weapon in flight towards the center of its reference point, and two launch carriages that hold the fiberglass containers with the missiles, to move and aim them towards the sector from which the threat is coming. These, which on some mounts are quadruple, are what collect the missiles stored in groups of four in two drums. The reloading process is automatic, and can reach an extraordinary rate of fire that can fire 10 missiles in a very short period of time in response to saturation attacks.

Finally, the acquisition unit includes Doppler surveillance radar that can detect objects within a radius of ten miles and incorporates a friend-enemy interrogator, a data processor for dealing with located traces and a presentation console for the commander of the vehicle.

FRENCH CAROL SYSTEM

L'Armée de Terre has ordered twenty Carol (Cabines Roland Tractées par un VLRA) systems to be quickly taken along with its rapid deployment units. Between 2001 and 2005, France should receive a further 72 similarly configured units.

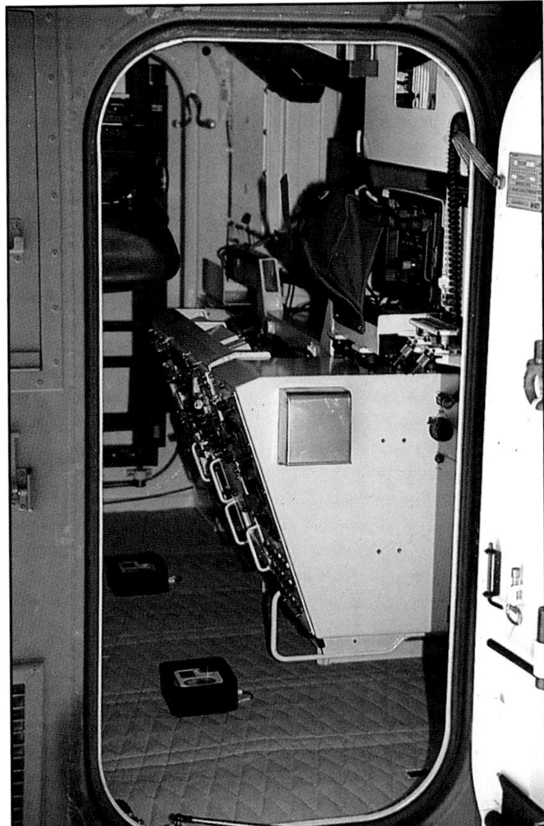

CONTROL UNIT

The inside of the Roland shelter contains a fire control panel from which two operators study the radar signals, oversee the performance of the anti-air system and take charge of guiding the missiles towards their targets. A large part of these procedures is semi-automatic.

ALL-WEATHER CAPABILITIES

The difference between 'all clear' Roland systems and those for all weather is that the latter include Thompson-CSF monopulse radar that is able to follow the target and a radio frequency transmitter at the rear of the missile to make sure that they meet at a certain point and produce the interception.

MISSILE CONTAINERS

The Roland on an AMX-30 includes two containers with four missiles ready to be fired and two quadruple drums, placed on top of the transport vehicle. These reload very quickly, permitting successive firing actions against saturation attacks.

ALL-TERRAIN MOBILITY

The advantage of the Roland fitted to the chassis of a tracked armored vehicle is that it can travel over any type of terrain thanks to its engine-driven power, provided by a turbo diesel engine of over 700 horsepower and situated at the rear. This is also place where there are a few boxes of spare parts and auxiliary tools.

HIGHLY MOBILE CATERPILLAR

The need to accompany armored forces and to defend them against the possibility of air attacks led to Rolands being installed on caterpillar vehicles, outstanding for their ability to move across any kind of land surface.

TECHNICAL CHARACTERISTICS ROLAND 3 ANTI-AIR MISSILE

COST OF FIRING UNIT IN DOLLARS:	
10 to 12,000,000 depending on configuration and transport vehicle	
DIMENSIONS OF THE MISSILE:	
Length	2.4 m
Diameter of the body	0.16 m
Span	0.5 m
DIMENSIONS OF THE CABIN:	
Length	4.7 m
Width	2.8 m
Height	1.7 m
WEIGHTS:	
Of the missile	75 kg
Of the explosive warhead	9.2 kg
Of the cabin with the firing position	8,300 kg
PROPULSION:	
SNPE rocket engine using two stage solid fuel	
FEATURES:	
Maximum range	5 miles
Speed	570 meters/s
Minimum interception height	less than 20 m
Lethal radius of explosive head	8 meters
Daytime use	yes
Nighttime use	yes
Manouverability	17 G'S

DETECTION AT LOW ALTITUDE

The Roland's firing position includes low flight radar detection that operates with Doppler pulses and is able to detect any aircraft within ten miles of the launcher. It is also capable of friend-enemy interrogation.

ARMORED CASEMATE

French And Spanish Roland are transported on AMX-30 combat tanks that are modified with armored casemates. The driver sits on the left and the commander of the vehicle on the right. These crewmembers have several periscopes for the protected observation of what is happening outside.

MAIN FIRE CONTROL POINT

The person in charge of the Roland fire unit (normally an officer) is the one who deals, via a console with a display screen, with the data received about the targets and the elements with which it could be defeated. They can fire simply by pressing a pedal designed for the purpose.

CAMOUFLAGE

On the top left hand side of the front section of the vehicle is a quadruple Weggman smoke launcher that can be used for either launching smoke bombs or anti-human grenades.

The French Crotale anti-air system has a very curious history. Its development had nothing to do with any kind of interest on the part of the French Armed Forces nor of the national industry in putting a new system onto the market, rather as a result of a British embargo on the sale of medium-range Bloodhound missiles to South Africa.

As it was unable to access that system, the South African government (with whom there were several limitations on arms exports due to the policy of apartheid that segregated the black community in that country) contacted the French Thompson-CSF and Matra companies in 1964 to persuade them to supply the weapons they needed. They required a new set of anti-air missiles that were suitable for tactical and strategic mobilization.

TOTAL MOBILITY

The chassis of the MLRS vehicle has been configured to move a Crotale NG anti-air defense module, a system noted for its excellent features in any weather conditions and for its ultra-compact design.

The decision to develop the Crotale

The South African investment (which covered 85 % of the funding of the project with the remaining 15 % supplied by the French government) meant the project could go ahead. In 1965, an unguided missile could already be fired to check the qualities of the launcher and meanwhile the project continued to develop the design of a system that could be used at very low altitudes even in an environment saturated by electronics.

Progress moves on swiftly

Two years later the first examples of guided missiles were fired, an incredibly short development period on behalf of the Matra company. In 1969, the complete system could be tested at the Landas site in France, and evaluations showed it has a 70 % chance of impact in the strictest operative conditions. The production line was quickly prepared. In 1971, the first elements of the 10 anti-air Cactus batteries were ready (the change of name owed itself to a political decision that wished to disassociate the weapon from the French model) for sale to South Africa. These included 60 launchers and 30 detection units fitted to 4x4 armored vehicles that could move them rapidly to their position of deployment.

This first order was followed by one from the French Air Force for 24 sets to equip 12 anti-air squadrons assigned to different air bases and civil airports that needed defending from possible air attacks, recently updated launchers with flat antennas on their radar and more efficient counter-measures. These sales inspired the investigation of a naval variant for the French Marine Nationale that was

EXTREMELY FAST MISSILE

The VT-1 is a new missile belonging to the Crotale NG family, shown here in cross-section to reveal the position of several components. Note that the warhead is at the front and the engine and flight components are at the rear. It was designed by the United States company Vought LTV.

complete by 1979. In 1980, the Shavine were sold to Saudi Arabia, made up of a caterpillar chassis based on n AMX-30C to move an improved version that used R460 missiles with a greater radius of influence and a better explosive charge.

Further orders arrived, including, for example, the development of all-weather variants equipped with an infrared system, as ordered by Saudi Arabia in 1980, and in 1985 design started on the NG (Nouvelle Generation) which, in principle, was designed to satisfy the requirements of the United States Army FAADS.LS-L-H program. Sales did not work out, but development continued. In the nineties several sales contracts were negotiated, and at the moment both the sheltered variant and the mobile one fitted to a MOWAG Piranha 10x10 wheeled armored vehicle or a United Defense tracked vehicle are available.

The Shahine, a unique proposition

In the mid seventies, the Crotale was a fixed use system that was moved by the four wheels built into the chassis to move the launchers; nowadays there are self-propelled variants on offer on caterpillar or wheeled chassis. These deployment limitations led the Saudi Arabian government to request the development or a more mobile version that could accompany armored troops, and that was also mechanized to protect its advance from aerial attacks from airplanes and helicopters. In 1975, at the Satory exhibition, the beginning of this project was announced.

RAPID RESPONSE CAPABILITY

The self-propelled variant of the Crotale (which combines an outstanding anti-air firepower) integrates eight missiles in immediate fire positions, three dimensional detection radar and pursuit radar.

With Saudi funding and using the chassis of an AMX-30C tank to allow for transportation over any kind of land surface, the Shahine was developed. It was an update of the Crotale, to which several major changes had been incorporated to improve its capabilities. This included a great deal of automation that minimized reaction time, six missiles ready to be fired instead of the four of the previous variant, a television camera coaxial to the surveillance radar, modernized features on the Doppler detection radar with a wider antenna that

TECHNICAL CHARACTERISTICS NG WITH VT1 MISSILES

COST IN DOLLARS:	**12,000,000 per fire unit**
DIMENSIONS OF THE MISSILE:	
Length	2.29 m
Diameter	0.165 m
Span	0.45 m
WEIGHTS:	
Of the missile	75 kg
Of the warhead	13.14 kg
PROPULSION:	
Solid fuel engine with low smoke emission	

FEATURES:	
Maximum range:	6 miles
Minimum effective range	320 miles
Operational height	From 15 to 6,000 m
Maximum speed	Mach 3.5
Daytime use	Yes
Nighttime use	Yes
Probability of strike	+90 %
Shots a minute	3-4
CREW:	
One operator and two assistants, one of whom drives the transport vehicle	

was transported on a different tank to increase its chances of survival, and the incorporation of a new missile called Sica or R460, measuring 3.12 meters in length, with a weight at launching of 100 kilograms, a range of 8.5 miles, a flight speed of Mach 2.5 and including a 15 kg warhead activated by an infrared proximity or contact fuse.

In service since 1980, this system was noted for its high cost, but also for its excellent range, capability of intercepting eight targets and short reaction time, all of which combined to make it far superior to anything similar at the time. They have recently been modernized by Thompson-CSF to maintain their excellent response time.

HELPING WITH WORK
The new generation Crotale has been automated so that only one man handles the console that controls assignation, launching and guidance of the missiles. The display elements are remarkably rational and totally digitized.

Well received

The features of the different versions of Crotale have led to the sales of more than 250 units, covering the initial 1,000 series, the 2000, completed in 1973, the 3,000 that was produced from 1975 onwards, the 4,000 created in 1983 and the 5,000 that was first produced in 1985. At the same time, the system has been copied for designing the Chinese FM-80 and the Korean Chun-Ma, which has the same guidance and detection elements.

Among the several users of the terrestrial version are Saudi Arabia (who acquired 48 launch units and 16 related to the Shahine), Bahrain, Chile (who operate four fire units), Egypt (who bought 24 launchers), the United Arab Emirates, France (who have recently acquired 12 NG sheltered systems for their rapid deployment forces), Holland (who assigned them to their Air Force), Libya (whose 27 systems are probably not operative after the

COMPACT AND EFFICIENT DESIGN
The French have a long tradition of developing anti-air systems, leading to a mobile sheltered design that encompasses all the elements that permit the autonomous functioning of aerial defense using the Crotale missile.

embargoes imposed on the country), Pakistan, and South Africa (who still has most of its 60 Cactus launchers in active service). Similarly, different naval mounts have been installed on ships belonging to Saudi Arabia, the United Arab Emirates, France, Oman and Taiwan.

Finland has recently joined the former (buying 20 NG systems that have been installed on XA-180 armored wheeled vehicles to improve their tactical mobility) and Greece (although that country already used a system produced locally under license called Apollo) acquired 11 NG sets in 1998, two of which will be used for defending naval bases and the rest by the Air Force for the protection of airport facilities.

IMMEDIATE AIR DEFENSE

French destroyers, and a large number of the ships used in the French Marine Nationale, rely on the Crotale for self-defense against airplane and helicopter attacks or for intercepting anti-ship missiles that approach them at sea level.

The New Generation

The latest development of the Crotale, being offered to the world under the name of NG, is a multi-mission air defense missile system. It is characterized by its easy integration and improved auto-coordination and mission flexibility thanks to the integration of detection, guidance and launch elements into a single module that is controlled by an operator via a completely automated and digitized console.

Optimization of characteristics

Four NG make up a fire group and work together via an automatic data exchange system that connects their processors. The nearest launcher takes control of the destruction of the target; therefore, it can be more easily integrated into a global air defense system. Similarly, its compact design means it can be transported in the hold of airplanes like the C-130 Hercules.

It is also worth mentioning that it has a better detection range, better information about threats, an IFF friend or foe identifier, the automatic locking of different targets, reduction in reaction and intervention time, better chances of survival, and a higher firing probability. All this means it can accompany mechanized units on their movements, defend vitally important positions or form part of zonal air defense forces against a wide variety of aerial targets, weapons launched from a safe distance and saturation attacks, even in adverse conditions of electronic warfare or where aggressive NBC elements are in use.

Constitutive elements

To maintain these advanced capabilities a system has been designed made up of an internal pedestal and a shelter or control installation that can also be placed inside the transport cabin of different types of armored

NG, THE NEWEST OF THE SAGA

The Crotale, whose design was initiated in the early sixties, has evolved in accordance to new threats, and has become a sophisticated anti-air defense system. Therefore, the French Thompson-CSF Airsys company, that markets it, is still receiving orders for its purchase.

SHELTERED VARIANT

The version of the Crotale NG installed in a shelter means it can be quickly towed to the deployment zone, while the detection elements can be folded up and fitted easily into transport aircraft holds.

vehicle. The former includes Doppler type detection radar operating on Band S, and includes an IFF antenna. It is capable of operating via pulse compression and frequency agility to fight against enemy ECCM electronics, and covers an area of up to 3 miles in height and 12.5 in range. It has TWT monopulse pursuit radar operating on band Q and with a range of up to 15 miles; a FLIR system based on a thermal camera that can choose between two sectors of a determined size and allows for the observation of objects situated up to 12 miles away; a daytime television camera that can see for 8 miles, and an infrared locator that allows for the following of the route of the missile.

To the former, which are responsible for the surveillance and pursuit, are added two side supports for up to eight sealed containers for placing VT1 missiles, originally made by the United States company Vought LTV.

NAVAL CARRIAGE

French aircraft carriers depend on octuple mounts of Crotale for their last line of air defense. They are positioned on the sides, from where they are able to face the threat from whatever angle it may approach.

These are noted for their flight speed of Mach 3.5, are capable of carrying out maneuvers of up to 35 g's and have an effective range that is 6 miles deep and 4 miles high. It is guided automatically, though the process is supervised by the systems operator via a CLOS (Command to Line Of Sight) system, designed

to take advantage of the simultaneous capacity of the different sensors; the guidance link is immune to counter-measures and includes a warhead that is activated by a proximity fuse via radio frequency with an 8 meter radius of effect; a typical interception would involve the destruction of a target situated 5 miles away in 10.3 seconds, which means it is possible to fight several targets in a very short space of time, and it is so maneuverable that it can also be used against missiles.

Several functions of the control facility have been updated, such that only 6 seconds are needed between detection and the moment of fire. The customized software takes charge of selecting the best guidance sensor according to the data that it receives. However, the operator can introduce changes if he feels it necessary, because his opinion is always valued above that of any of the automated components.

ALL-WEATHER CAPABILITY

By day or night, or in bad weather, the surveillance and pursuit elements integrated between the missile containers can be used for all the relevant operations without the system suffering any kind of restriction.

RADIO DEPLOYMENT

The need to move troops and materials quickly to places far away from base to participate in all types of mission (whether for combat or peacekeeping) demands a high degree of mobility from all of the systems for transportation by air or by sea.

DEPLOYED BY THE CANADIANS

The Canadian Army received 36 ADATS systems that have been installed on M-113 armored caterpillar vehicles to give them the mobility they need to advance with and support mechanized units and to provide anti-air and anti-tank cover.

The arms race that began in the seventies resulted in numerous weapon systems and equipment of all kinds being introduced during the early years of the nineteen eighties. Among these was the ADATS system (Air-Defense Anti-Tank System). This was born out of the need to confront two of the most feared threats on the modern combat fields; the tank and ground attack aircraft, which were strong enough to stand off or halt the progress of advancing terrestrial units.

Different concept

This double threat could approach, then, either by land or from the air, which was the ideology behind the studies that were carried out by engineers at the Swiss firm Oerlikon-Bührle from 1973, in which the sales and market shares potential of a system capable of destroying both airborne and terrestrial targets was analyzed.

Shared work

To carry out the project, in 1979 they joined another company, Martin Marietta in the United States, who already had ample experience in the development of missiles and their associated guidance systems. They provided 150 million dollars to produce two mobile prototypes, which were to be installed on a caterpillar vehicle belonging to the M-113 family and on a wheeled Cadillac Gage V-300 Commando armored vehicle, and a third sheltered version. The evaluation tests on the ballistic missile took place at the Martin Marietta Test Center at Orlando, Florida in November 1980. In April 1981, the

AHEAD AIR DEFENSE SYSTEM

The combination of the SkyShield 35 air defense system with ADATS launchers increases its overall capacity and allows for its use with missiles or guns against threats according to their distance, altitude and capacity of reaction.

weapon was guided by a laser illumination beam during tests at the missile firing range at White Sands in New Mexico. Midway through 1982 the first missile was fired from one of the prototypes mounted on a vehicle, and towards the end of 1983 development was complete after 39 trial missiles had been fired.

The good results of the initial evaluation process, and the clear potential of the system, were followed by its promotion, which began in 1984 with the presentation of one unit to the Swiss army for evaluation and possible purchase.

The international marketing campaign begins

Although the operative prototype had already been taken to Le Bourget aeronautical show so that potential clients could get to know it, the first significant reply did not come until 1986 when, after nearly a year of evaluation, the United States Army selected it for its FAAD-LOS-FH (Forward Area Air Defense Line Of Sight, Forward Heavy) program, for which it named it MIM-146.

At the same time, the system came under the Canadian Armed Forces' consideration, who became its first user after buying 36 units that were to be installed on caterpillar vehicles of the M-113 series and by Canada's Oerlikon Aerospace, with the aim of satisfying LLADS (Low Level Air Defense System) requirements. The last one was received in 1994, just two years after the United States cancelled its FAAD requirement in which it had used four prototype systems and carried out more than two hundred launches to verify the real potential of the weapon.

TRIED IN THE DESERT

Saudi and Kuwaiti requirements for the purchase of a highly capable anti-air system led to the adaptation of a Canadian ADATS mount to complement evaluations in the Middle East. It showed itself to be capable of detecting and neutralizing different kinds of targets in desert conditions.

OPERATED BY THE THAIS

The Thai Air Force received a sheltered version of ADATS that does not include search radar and is used in connection with Skyguard fire direction for the protection of Thai air bases from the possibility of attack.

That same year a sales contract was signed with Thailand (after five years of negotiations) for between 10 and 20 sheltered fire units without radar that are now used by that country's Air Force. They are associated to Skyguard fire direction control that, with its surveillance and pursuit radar, provide them with the data related to the position of the target. To find new potential clients (who could include Saudi Arabia, Turkey and Kuwait), a Mk 2 variant was developed, that includes several improvements such as a new CCD (Charge-Coupled Device) television module in the electro-optic system, a 60 % smaller radar electronics container, some new multi-purpose consoles that can alternate information from the radar or from the electro-optic system, an integrated C31 system, a new hydraulic power unit and a redesigned air conditioning system, while the missile has been adapted to modern day defense requirements.

Capacity for use

The initial requirements of ADATS included its use as an anti-air weapon and also against tanks, while the designers even managed to add a method for using the missile as an air-surface weapon launched from fighter-bombers. Although the missile showed in trials that it was capable of impacting with and destroying medium tanks within a radius of five miles thanks to its explosive charge that can almost penetrate a meter of armor, the reality is that the cost of the associated system and missile have led to it only being used for anti-air purposes. At any given moment, it can be modified for self-defense purposes as protection against advancing armored forces.

EIGHT MISSILES READY TO BE FIRED

After detecting the intruder and its initial pursuit, an ADATS missile is directed towards it at a speed of Mach 3. It can strike it within a radius of 6 miles. Here we see the weapon in the moment that it leaves the launcher.

MODULAR SYSTEM

ADATS was originally designed as a modular system with a strong capacity for adaptation to the different kinds of support that move it. These include the palletized mount pictured here that facilitates its rapid transport by any terrestrial, aerial or naval medium.

Highly capable of neutralizing the enemy

The modular design of the system (from the beginning, intended to be of a compact size for its easy transport on a variety of different transports) includes a turret for providing its containers with mobility. They can be situated at an angle of depression of −10° or of 90° elevation and it can move them at angular speeds of two radians a second in horizontal and of one in elevation. It includes surveillance radar that can locate up to 10 targets within a radius of some 15 miles, and also a built-in IFF Identification of Friend of Foe system; two control consoles handled by their respective operators on which the radar data is displayed and the electro-optic pointing element; the calculation elements related to the pointing unit that is mounted on a stabilized platform and encompasses an infrared system associated to a television camera, a Yag-Neodymic laser for measuring distance, an infrared detector that measures the diversions of the missile in relation to the line of sight and a codified CO_2 laser that serves to direct the missiles towards the impact point.

The latter are come in containers where they can be stored for 15 years without needing checking or maintenance. They include an element that receives laser signals from the launcher for guiding it, the elements

TECHNICAL CHARACTERISTICS ADATS ANTI-AIR ANTI-TANK MISSILE

COST IN DOLLARS:	Unknown
DIMENSIONS OF THE MISSILE:	
Length	2.05 m
Diameter of the body	0.15 m
WEIGHTS:	
Of the missile in its container	67 kg
Of the missile	51.4 kg
Of the warhead	12.5 kg
Of the palletized firing mount	3,480 kg
Of the sheltered firing mount or for installation in vehicles	4,500 kg
PROPULSION:	
Rocket engine with solid fuel	
FEATURES:	
Maximum range	6 miles
Maximum operational height	4 miles
Daytime use	Yes
Nocturnal use	Yes
Maximum speed	Mach 3+
Maneuverability	60 g's
CREW:	
Two operators in the transport vehicle	

that activate the explosive head through either impact or proximity and an infrared transmitter that means its route can be followed from the control unit.

Its strange guidance system (which combines radio control in the line of sight in the first phase with the use of the laser beam guidance of the second) makes it possible for just one fire unit to confront four airplanes at the same time, due to the great speed of the missiles and the remarkable reaction times of a system designed to work without restrictions even in conditions saturated by countermeasures. Similarly, it is significant that in three or four seconds (the time that it takes for the gunpowder in the twin base of the engine to burn) the missile reaches its maximum velocity. Corrections of the route are carried out by four small fins built into the rear section and taking automatically sent orders from the laser beam director.

Configuration of point defense

The changes in acquisition policies that were suffered by the rapid reduction in defense budgets and the significant lessening of real threats of war have led the commercial department of Oerlikon Contraves AG (the company dealing with its promotion) to make

OUTSTANDING FEATURES

The ADATS missile system stands out above others for its speed in excess of Mach 3, the power of its combat charge, which includes an explosive charge of 12.5 kg, and its unusual CO2 laser guidance system, making it a very accurate weapon at any altitude.

PALLETIZED MODULE

The latest version of ADATS, and the lightest (now measuring nearly 1,000 kilograms less than that used on vehicles), is this highly mobile pallet that includes an electro-optic guidance unit in the center and the eight missile containers on either side.

COMPACT AND DUAL PURPOSE

The ADATS system is noted for its eight missiles positioned in two quadruple launch containers that can destroy both land and air targets. These targets are detected by the identification systems built into its very mount.

it part of the AHEAD/Sky Shield air defense system in the hope of making some kind of sale.

Basically, this is made up of different sheltered modules, easily transported by air or in medium capacity trucks to take them to the positions near those that need defending. Its different components are overlapped at the point of action. The basic element is the SkyShield fire control unit that includes Doppler pulse radar for detecting all kinds of airborne objectives (including missiles of 0.1 meters squared in a 6 mile radius), monopulse radar for pursuit, an infrared television system and a CO2 distance measurement laser, mounted on a sensor unit that weighs 2.200 kg and can use its electro-optic system passively to avoid being located. Two single barrel 35/1000 guns are connected to the former, and do not need operators because they are remote controlled. They have a firing rate of 1,000 rounds a minute that can be maintained for several bursts of fire thanks to its 228 rounds.

DIRECTION OF FIRE

In the AHEAD configuration, the SkyShield fire control unit is fundamental for detecting threats and transfers the details of its position to the ADATS launchers that will be able, once they have been activated by their operators, to intercept within a radius of 6 miles.

The United States Patriot anti-air system became very famous as a result of the Gulf War where it was used to protect different places in Saudi Arabia, Israel and Turkey from the attacks of ballistic Scud missiles.

Saddam Hussein's threats came true with the launching of several Scud missiles against important Israeli towns, and it was believed that those missiles could contain aggressive biological and chemical elements in their warheads. Therefore, 160 Patriot missiles were launched to attempt to take them out in the air, or explode the warheads and change their flight paths. General public opinion pointed towards a success but, as far as the number used in comparison to results obtained was concerned, it could be said that they were a military failure.

EXCELLENT FIRE CAPACITY

Eight towable launchers for MIM-104 missiles are included in each Patriot battery, so that it has up to 32 missiles to confront an intense threat. It has a very high capacity for reaction, and only 15 seconds are needed to reload the four containers.

Development of a new anti-air missile system

The continued increase in air threats led the strategists at the Pentagon to begin a new medium-long-range anti-air missile project that could complement and eventually substitute the Hawk and the Nike Hercules. The first studies began in 1963. These original scale models, part of the United States Army FABMDS (Field Army Ballistic Missile Defense System) program, were used some years later as the basis of a new program, AADS (Army Air Defense System) that was developed in the seventies.

Defense of ground forces

As a result of the aforementioned, the SAM-D (Surface to Air Missile Development) was created; the development of which began

THE GREATEST MOBILITY

All Patriot systems are situated in shelters or wheeled platforms that get their mobility from different types of truck, amongst which are included these four axles for towing the missile launcher-erectors.

in 1966 based on contracts signed between Hughes Aircraft/McDonnell Douglas, RCA/ Beech and Raytheon Martin Marietta. The definite contract was finally signed in May 1967 with Raytheon, who were assigned the role of principal contractor. The idea of new multi-purpose radar and a missile, called MIM-104[a], that used advanced technology, became a reality extremely quickly, and four years later a system was produced that surpassed initial expectations. The tests that followed lasted up until 1973.

The lack of funding delayed the start of production until 1979, during which time the manufacturer modified and improved the original design. A 57.8 million dollar contract was finally signed, accepting the challenge of producing 15 Anti-Air Defense groups. The first of these was ready for active use in September 1984, and was sent to the 7th Army posted in Germany.

While the first systems were entering service, improvements were already being made, aiming at what was then known as PAC-1 (Patriot Advanced Capability), which affected the algorithms of the search and pursuit software. The first trials were carried out in 1986. A short while later the PAC-2 was conceived, including an improved MIM-104C missile and changes in the radar for detecting smaller targets, and it was ready for use in the 1990-1991 Gulf War. MIM-104D missiles came later as part of the GEM (Guidance Enhanced Missile), and in 1994 Raytheon and Lockheed Martin Vought Systems received the development contract for the PAC-3, which included newer a smaller missiles (in an old 4 missile container that now holds 16) and improved capabilities. In January 1999, these companies established a contract with the German company Daimler Chrysler Aerospace for production in Europe. This was particularly efficient at destroying all kinds of tactical and cruise missiles.

MULTI-PURPOSE RADAR

The AN/MPQ-53 system is made up of a shelter and Doppler pulse radar used for air surveillance, target detection, pursuit of those that pose the greatest threat, identification of enemies by means of a built-in IFF, illumination of the missiles towards the target and guidance throughout the final stage of flight.

FIRE POINTING POSITION

A watertight cabin protected against NBC, which is moved on a 6x6 truck, encompasses the different elements of the direction point that includes the central elements for controlling the whole system and deciding exactly which objectives to attack and at what time.

Its high cost has not hindered sales

Parallel to its entry into service, the Armies and Air Forces of several different countries expressed an interest in buying Patriot systems to greatly improve their capabilities with respect to anti-air defense. The US Army has contracted 104 fire units and some 6,000 missiles. At the same time, 56 launchers and 3,000 missiles have been made for export to other countries. These include Germany, who uses the system in the Air Force and hopes to modernize them into the PAC-3 variant over the next few years; Saudi Arabia, who have 21 systems; Holland, who deploy four fire units; Israel, who received a large percentage of theirs as compensation for staying inactive despite the Iraqi provocation of 1991; Japan, where the Mitsubishi Corporation produces it under license; Kuwait, who received it after the extensive military acquisitions plan that followed the experience of the Gulf War in which its semi-

TECHNICAL CHARACTERISTICS MM-104D MISSILE

COST IN DOLLARS:	**1,300,000,000 per battery**	Maximum radar range	15 miles
DIMENSIONS OF THE MISSILE:		Maximum missile speed	From Mach 2 to Mach 5
Length	5.2 m	Speed of interception	From 1 to 2.5 miles/sec
Diameter	0.41 m	Daytime use	Yes
WEIGHTS:	**914 kg**	Nighttime use	Yes
Of the missil	914 kg	Effectiveness	96 % with the US Army
Of the explosive warhead	90 kg	Shots per minute	Multiple through use of several launchers
Of the solid fuel	600 kg		
FEATURES:		**CREW:**	
Maximum missile range	55 miles	One hundred men per battery	

mercenary armies showed what little use they could serve against the Iraqi advance.

To these, Taiwan can be added, who acquired the MADS (Modified Air Defense System) or Patriot T, which consists of an improved version of the original assigned to protect the heavily populated Taipei; the first units were received in 1997. A large amount of this technology has been exploited to produce the Tien Kung system, a Taiwanese design. In early 1999, Egypt expressed an interest in a PAC-3 battery with 32 missiles that would cost about 1,300 million dollars, and there is a chance that they might ask for two more. Several other countries look likely to be added to the list of purchasers, including Spain and Turkey.

Operational conception

The Patriot was originally conceived as a zonal air defense system that would be moved on towed vehicles or heavy trucks that could take it to the deployment position and establish batteries just a matter of hours after the anti-air mission has begun.

Theoretically exceptional capabilities

The need to operate against large formations of aircraft in an area where electronic countermeasures are being used intensively was why the system has such a high fire capacity that means it can be used for multiple and simultaneous interceptions. For this reason, the system was designed to be so automated.

DISTANT AIR ALERT

Although the Patriot itself has the capacity to detect aircraft within a radius of 105 miles, it is normally complemented by mobile radar that can detect the presence of enemy craft within a radius of 250 miles and then transfers additional information to the missile system. (left photograph).

RANGE OF 55 MILES

The missiles, stored in aluminum containers until they are launched, are able to reach targets within a radius of 55 miles thanks to their high speed and the power of their explosive charge that expels large fragments that can damage both missiles in flight and every type of aircraft (right photograph).

To serve this purpose, each battery was organized according to different elements that work together to achieve the best overall effect. The nerve center of the Patriot's capabilities is the mobile unit that contains the AN/MSQ-104 ECS Engagement Control Station, which directs fire. It is made up of a 6.76 meter aluminum cabin, mounted on the trailer of a 6x6 AM General M-816 tactical truck, where two operators work with display screens related to the monitoring of the battle, establishment of safe passages for its own aircraft and controlling the rules of the encounter. This task is supported by a powerful processing unit and HF and VHF communication equipment to provide connection with other elements of the group.

Complementary to this is the AN/MPQ-53 multi-purpose radar with Doppler pulses that take charge of surveillance, detection, pursuit, identification, illumination and missile guidance, confronting any electronic countermeasures. It includes an air conditioned and electronically isolated cabin that incorporates a large, flat antenna the operates on band C at the front, including some 5,000 phase elements and that can control the threat of the 100 potentially most dangerous targets. It is connected to the direction center via a fiber optic cable.

This also normally links with eight M-901 launchers (in reality up to 16 can be controlled by one radar), that are situated on an extremely large towed vehicle. This is stabilized by four hydraulically conditioned jacks, which have circular and mobile bases that allow the four containers that they transport to be pointed in the appropriate direction before the missiles are fired. Next to the former, its own electricity generator is towed. Other elements of the system include: the ICC shelter, assigned to each battalion so that it can control up to six fire units acting as an Information and Coordination Center; the AN/MSQ-24 electric current generator; trucks with spare missiles and movable AMG (Antenna Mast Group) antennas and CRG (Communications Relay Group).

FUTURE PROGRAM

The US Army is working on the THAAD (Theater High Altitude Area Defense) system based on solid state band X radar that would control the launching of the Patriot to intercept enemy ballistic missiles.

OPERATION DESERT STORM

During the Desert Storm operation against Iraq, large numbers of Patriot batteries were deployed in Saudi Arabia and Israel, where they defended the most important facilities and cities from the attacks of Scud missiles. To do this, they are camouflaged with the same colors as the actual desert.

Strange missile with respect to possibilities

The MIM-104 is a missile that is transported in a watertight aluminum container-launcher that includes two rails on the inside along which it slides when it is fired. It is configured by a front radome made of silica ceramic, the guidance section, the fuse, two small relay antennas, the warhead (which, in its PAC-2 variant generates 45 gram fragments when it explodes), the gyroscope, the inertial, a large container with the propeller, the rocket engine and the guidance fins. All these elements have been gradually improved with each new variant up until the latest PAC-3 missiles.

However, one outstanding characteristic is the way it flies towards its target under guidance of its inertial unit and is propelled by its hydrooxylpolybutadion solid fuel rocket engine that burns for 12 seconds and provides a speed measured at anywhere between Mach 2 and Mach 5. Guidance is carried out in its terminal phase by its semi-active TVM (Track-Via-Missile) radar system that acts for the last 10 seconds of the flight of the missile. This means several missiles can be fired at detected targets and can be guided in sequence through the last seconds of its flight towards interception.

GROUP OF ANTENNAS

Each Patriot battery includes its group of UHF directional antennas that facilitate links between the different sections that make it up, and also connect, through relay stations, to the different direction points of other batteries.

PROVEN IN COMBAT

The characteristics of the Patriot system have been demonstrated positively in combat, although some reports do suggest a certain lack of effectiveness. Consequently, new systems such GEM guidance, PAC-3 missiles and improved radar have been added.

Bofors 40-millimeter guns have played a significant role in the anti-air artillery of several countries for seven decades, and they are expected to continue doing so for several more. This is mainly possible thanks to its extraordinary robustness, the fact that its caliber has made it possible to produce increasingly more advanced and capable munitions, and it can be made more accurate by incorporating new aiming elements that benefit from the latest electronic and optic technology.

These capabilities (which are well known to those combat pilots who have had to execute missions in areas where these powerful and accurate weapons are operating) have been developed for the making of terrestrial and naval varieties that are used by a multitude of different countries in every corner of the globe.

EXTENSIVE NAVAL USE

Simple Bofors 40/70 mounts are used on different kinds of ship (from patrol boats to frigates) as a complement to their anti-air/anti-missile defenses and even as an auxiliary measure against surface targets.

SIMPLE TO USE

The four operators of a 40/70 anti-air weapon only have to follow the orders that they receive from the fire direction center to shoot towards the area that contains the predicted hazard, and to fire the mount and reload it with the combs that take up the munitions.

Designed to overcome the deficiencies of war

The Swedish Armed Forces requested the Bofors company, based in their country, to develop a medium caliber anti-air weapon that would satisfy the needs of self-defense that had become apparent with the disappointing results of the 20 mm one that had been used in the First World War.

The prototype is prepared

The first test unit appeared in 1930 and its development was completed four years later.

SPANISH ARTILLERY

The Spanish Ejército de Tierra still uses nearly two hundred 40/70 millimeter weapons for defending areas prone to all kinds of air attack, for which purpose they have often been updated, and they should now be able to stay operative until 2010.

Production began for the M1934 weapon (which would be revised two years later and called the M1936), and thousands of units were built in Sweden and also Hungary, Poland and Norway, countries that had received production licenses. With the advent of the Second World War, huge quantities were exported to Australia, Canada, the United States and the United Kingdom. This 40-mm mount is used in a wide variety of configurations that include an aerial carriage used for hunting tanks and perforating them with its powerful munitions.

To improve the capacity of the L/60 mount, which was still being used by ten countries in the late eighties, development work began in 1945 on a new family of munitions that needed the original 60 caliber barrel to be changed for a 70 mm one to increase the initial speed and therefore satisfy new requirements. This was how the Bofors L/70 came to be, shooting its missiles at 1,025 meters a second, reducing the flight time up until reaching the point of impact and able to fire at a rate of 240 rounds a minute, which increased the possibility of striking the target.

HALF A CENTURY OF DEFENSE

Towed Bofors 40/70 mounts have been in service for over half a century. Their different configurations have developed over the years in relation to the changing needs of its users and different types of threat.

Used worldwide

From 1947, the year in which this model was first introduced to the market, the production license has been passed on to Brazil, Spain, Italy, India, the United Kingdom and to other countries that have produced it by the thousand, with orders arriving from all kinds of other countries.

The initial design was substituted by the improved L/70 model A mount that needed an electric power source for its movements and the L/70 model B that incorporated its own

generator at the front of the gun carriage. Midway through the nineteen sixties the BOFI variant was introduced, incorporating its own centralized fire control system with radar and measurement laser that improved its capacity for neutralizing airplanes flying at faster speeds.

This concept was expanded with the introduction of the TRINITY model, conceived as an alternative to the anti-air mounts configured for short-range missiles. Both the weapon and the direction system needed revising to improve their general performance capabilities, which now reach firing rates of 330 rounds a minute. In the same way, the fuses of the projectiles could be individually programmed in the moment of fire to increase their potential, which paved the way for a new generation of munitions, known as 3P (Prefragmented, Programmable and Proximity), and which used, among others, the Swedish CV 9040 IFV built onto a CV90 armored caterpillar vehicle.

Different updating processes

The need to maintain these pieces of artillery in a state that could obtain the maximum efficiency on the missions demanded of them has meant different countries, according to their particular capabilities, undergoing different updating and modification programs.

PROVEN VERSATILITY

The many years that the 40 millimeter Bofors has been in service have not been interpreted as a negative point by the Swedish Army, who decide to incorporate the weapon on the CV9040 anti-air combat vehicle, using a fire control radar.

Improved pointing systems

Among the different updating options for 40/70 mm artillery weapons that were proposed by the different companies is that of the Swedish Saab Aerospace company, part of the Saab Dynamics Group AB, which is based in Jönköping. The idea developed by this industrial group, also applicable to other anti-air artillery models, such as Soviet 23 and 57 mm

UPGRADED CARRIAGE

Taking the 40/70 mount as a base, an automatically fed Breda feeder has been designed that only needs two operators to control it and can maintain firing rates of over 300 rounds a minute.

guns (which have been used in their thousands by Eastern Bloc and Middle Eastern countries), is based around the idea of installing an integrated fire control system known by the acronym of LVS.

Its adoption converts it into a highly efficient and autonomous anti-air weapon against helicopters and airplanes. They can even fire at night thanks to an infrared sight, and the training for its operators is also improved through the inclusion of autotesting elements that provide continuous monitoring of the technical status of the system. The elements that make up the LVS system are a pursuit and target locking unit associated to one of the pointing axis of the main barrel; the control and processing unit; the control and display unit; an element that joins a gyroscope and a pendulum to calculate the exact position of the piece; the azimuth transducer, and the simulation unit used for training.

These elements, validated in 1994 by the Swedish Anti-Air School in a real firing exercise, greatly increase the probability of striking the target. A target was used for these tests that moved at 150 meters a second and was situated at a distance of between 1,500 and 3,000 meters from the weapon, at which ten rounds of munitions were fired. The results of these tests showed that 37 impacts were made in an area of 4 meters around the target, 20 in an area of 8 m, 8 in that of 12 m and just one completely missed the area of influence of the proximity fuses that activate the explosive charge and the multiple warhead in the moment that it crosses its flight path.

SWEDISH UPDATE

The SAAB Aerospace company proposes an updated single-barrel Bofors 40/70 mount with a fire control system, called LVS, that considerably increases the efficiency of the system to a level that 80% of the projectiles are fired within 8 meters of the target.

Spanish solution for the future

The Spanish Ejército de Tierra includes two hundred and fifty 40/70-mm guns in its anti-air artillery units. These were made by S.A. Placencia de las Armas (SAPA) between 1956 and 1962. To prepare them for the future a modernization program began in 1982, designed by the Jefatura de Artillería (Artillery Office) and with the intention of

TECHNICAL CHARACTERISTICS BOFORS SAK 40L/70-350 GUN

COST IN DOLLARS:	750,000
CALIBER:	40 mm
DIMENSIONS:	
Length in firing position	3.13 m
Length of barrel	2.8 m
Length of internal grooves	2.42 m
WEIGHTS:	
Total in combat order	2,790 kg
Of the tube	163 kg
Of the munitions	2.4 kg of HCHE and PFHE, 2.5 kg of HE-T

FEATURES:	
Effective range for aerial targets	2.5 miles
Effective range against surface targets	5.2 miles
Maximum range	7.8 miles
Rate of fire	300 shots a minute
Angle of elevation depression	+85°/-9°
Initial speed of projectiles	1,000 meters/second
Probability of strike	+30 %
CREW:	4 men

upgrading 164 guns at a rate of 24 weapons a year.

The process was expected to last seven years, ready to take on the nineties with a completely rejuvenated set of weapons with new immediate anti-air capabilities. At the same time, it is able to deal with every kind of surface target.

The cost of this process is predicted to fall at around 1,600 million pesetas (11 million dollars), which have been invested in substituting the valve amplifiers with transistor ones; the addition of a light electrogenics group that includes a two cylinder gas engine to generate the electrical energy needed for movement; the installation of a Contraves initial speed measurement system subcontracted to the Spanish company INISEL, and the preparation of the weapons for using dual-purpose proximity munitions against aerial and surface targets.

The increased firing rate, which has moved up to 300 shots a minute, and the arrival of new PFHE prefragmented rounds with programmable proximity fuses (that activate the detonation of the Octol HMX/TNT to impulse 650 wolfram spheres) means that they have enough surveillance and fire capacity to remain active well into the first decade of the 21st century.

ITALIAN ANTI-MISSILE MOUNT

The features of the 40/70 gun, and in particular that with prefragmented munitions associated to a proximity fuse, take into account the design of the Italian Dardo naval carriage, which is made up of a two barrel mount and fire control radar.

COMBINATION OF SECTORS

To guarantee the defense of a specific point against air attacks the different artillery mounts are situated in a way that covers the approach sectors and, in particular, those areas that the aircraft can use to advance without being detected.

The need to counter the air threat and, in particular, those of two specialized planes in attacking terrestrial positions, led to the development of different self defense systems. These include rapid-fire anti-air guns; weapons that are able to create fire at rates around 1,000 rounds a minute.

Using the right munitions, with the support of the most modern fire control elements and operated by well-trained crews, these guns are able to bring down the most sophisticated airplanes or prevent them from firing at the desired position. Among the companies that produce this kind of system is one of particular note, the Swiss Oerlikon, who has sold its specialized products in all corners of the globe.

Consolidated tradition

The current industrial group Oerlikon Contraves, with its headquarters in Zurich and delegations in other countries, originated from the merging of several Swiss companies in 1924. Since then, it has not stopped growing with respect to its industrial and human potential, and has developed a wide range of products, in particular guns and anti-air fire control.

HIGH FIREPOWER

The Oerlikon 35/1,000 gun has been designed to be remote controlled without the need of operators. It includes 226 AHEAD cartridges for carrying out a number of firing actions.

INTEGRATED SYSTEM

The Air Defense of sensitive positions requires different anti-air systems to gain the maximum effect. A good example of this is the Oerlikon GDF 35/90-gun working together with Skyguard direction and Aspide launchers.

Complementary projects

The lightest range of Oerlikon products includes the widely used 20 mm mount that is available with one or two barrels and of which several different lengths of gun and levels of firepower have appeared on the market. The most significant (that can also be used against surface targets) is the 20/120 model. It is made up of a carriage that can be operated by two men, fed by loaders for 20 rounds or

drums for 50, and able to fire, theoretically, at 600 rounds a minute. It can take out helicopters in stationary flight and planes flying at low speeds.

The Diana is stronger; it uses two 25-mm KBB barrels on a towed mount of 3,800 kilograms. It integrates a Gun-King optic aiming system and its associated laser, and each of its barrels is able to fire projectiles at a rate of 800 a minute, and they only need 2.4 seconds to travel a distance of 1.25 miles. Based on a similar gun, the KBA, it has developed the 25 Sidam turret that the Italian Army uses on M-113 armored caterpillar vehicles. Also worthy of mention is its quadruple mount of 25-mm weapons connected to a modern pointing system.

Twin 35-millimeter mount

The prototype was completed in1959 and, since the beginning of the seventies, began to be sold as the GDF-001. Its production license was passed on to Japan, and 1,800 units were sold to a total of 25 countries, which include Argentina, Ecuador, Spain, Greece and Turkey, countries that use both the terrestrial and naval varieties.

This mount (which is now available in the GDF-005 terrestrial configuration) is made up of two 35 millimeter internal diameter barrels that are 90 caliber in length and capable of firing at an overall rate of 1,200 projectiles per minute. This rate is increased if it is complemented by an efficient range of about 2.5 miles.

DESTRUCTIVE CAPACITY

AHEAD munitions have been equipped with a specific explosive charge, it fires 152 sub-projectiles at the target very accurately with respect to the point of activation. Its effect on missiles can be appreciated in this photograph.

DUAL USE

The capacity of the single barrel 20-millimeter Oerlikon mount for adapting its elevation means that it can also be used against surface targets. This characteristic increases the self-defense capabilities of the artillerymen that control it.

The terrestrial mount, which measures 7.87 meters when in operation and weighs 6 tons (the naval ones reach 6.52 tons in the GDM-A variant and 8 in the GDM-C) includes two automatic loaders with capacity for 56 destructive, tracer or perforating shots. Two men are needed to operate it, because the mount is automated and includes an associated electrogenic group for providing autonomous movement.

It is worth pointing out that the useful life of the barrels depends substantially on its use. Therefore, if it fires 3 or 4 projectiles per barrel at intervals of 60 seconds, these will last for 5,000 rounds. If it is used in series of 4 rounds of 8 shots with two seconds between each of them, its useful life is 2,500, and if 4 rounds of 50 rounds with two seconds between each are fired, it would need replacing after 185.

Finding new markets

The need to confront the continuous challenges of the market have caused the technicians at Oerlikon to design a new range of artillery mounts that combine great lightness, a notable level of firepower, the capacity of operating without needing the continued presence of a crew to oversee progress and a the need for a low amount of maintenance work. All this suggests that they will be adopted by several countries that need to increase their current capabilities to face up to the changing threats of the new millenium.

Revolver gun

It is made up of a mobile and air transportable mount built on a 35/1000 gun that works on gas intake and includes a revolver breech that holds four rounds. This substantially increases the rate of fire to 1,000 per minute. With a weight of 450 kg and a length of 4,110 meters, this weapon can be installed on a terrestrial carriage that weighs 3.5 tons. It can fire at 20 targets thanks to the 228 rounds situated in its stores and it can accelerate at two radians a second in transverse speed and acceleration. It receives the necessary electric power from a 1.5 kW generator.

COMPACT FIRE DIRECTION

The Contraves Skyguard is a fire direction system built on a highly mobile shelter that includes radar detection, pursuit radar, television cameras and display consoles, elements that are used for guiding the fire of the anti-air guns accurately.

What really give this system its artillery power are the AHEAD munitions. These include a 750-gram projectile equipped with a fuse that is programmable with regards to activation time with a system installed at the mouth of the gun. Its accuracy is as low as a thousandth of a second, which means it can strike a target area of one meter.

Fire control systems

The intrinsic fire capacity of this artillery weapon is made more effective by the use of sophisticated pointing elements. These go from advanced optic pointing systems, which combine an all-weather capacity with laser measurers that work out the exact distance of the target, to sophisticated fire control that combines detection and surveillance radar, television cameras thermographs and other elements.

SINGLE AND TWIN BARREL

The simple 20-millimeter Oerlikon mounts, of which so many variants are available both with one and with two barrels, are weapons used for immediate defense against slow flying aircraft or helicopters.

The widely used Skyguard is particularly outstanding, consisting of a small towed shelter with display consoles and the two operators that control them. More than two hundred have been made since 1979. These receive the information (which they pass on to the pieces via a relay cable) that comes from the sensors, made up of detection radar on bands I/J, pursuit radar and television cameras. Its range is about 12 miles and its resolution is 160 meters.

Skyshield is far more modern and capable, in which the operators sit in a specialized shelter and control the direction of fire and the associated missile launchers and artillery. Among its more outstanding characteristics are its weight of 3.2 tons, its length of 3 meters and transport height of 2.22. It is as accurate as 0.1 meters and includes Doppler pulse pursuit radar that operates on band X and can detect objectives on a surface of 0.1 m2 at 6 miles, pursuit radar, a laser yag-neodimio distance meter that emits 3 milliradian pulses and an associated television or infrared camera. All these elements permit automatic and effective aiming of the associated guns.

INITIAL SPEED MEASURERS

What may seem like compensators fitted to the normal barrels of the carriage are initial speed measurers that include circular rings that detect the speed of the projectile and carry out any necessary modifications to guarantee the most accurate aiming.

VERY LONG BARRELS

The 35/90 mount includes two high quality barrels that are noted for their length of 3.15 meters. These increase the initial speed of the projectiles and also improve the accuracy in the longest range, which is about 2.5 miles.

VERY CAPABLE PROJECTILES

In this version of the twin barrel mount, a rear element has been integrated that automatically reloads the groups of projectiles into the front or main loader. This increases the firing speed and generates more sustained fire that can deal with multiple targets.

ELECTROGENIC GROUP

GDF-005 mounts integrate low power diesel engines at the front that move an electrogenic group that generates enough energy for the automatic functioning of the weapon. This configuration is more compact and requires fewer vehicles to transport it.

OPERATOR'S POSITION

At the back of the mount, there is a position for the operator of the unit. He has an optic sight for aiming and a laser telemeter with which he can act as an auxiliary to firing should there be a problem with the automatic orders, or if this facility is switched off.

INTEGRATED SUPPORTS

To make the weapon more stable when it is in the firing position, the four wheels used for mobilization can be folded away and replaced by large jacks that stabilize the platform on any kind of surface.

TOWING HOOK

The 35-millimeter Oerlikon artillery mount is transported by a heavy triple axis, 10-ton truck. It is towed by means of a hook situated at the rear end of the carriage.

ELEVATION MECHANISMS

The mass of the breeches and barrels is moved vertically by robust elevation mechanisms that work electro-hydraulically. They can be activated manually in case of a power failure.

TECHNICAL CHARACTERISTICS OERLIKON 35/90 GDF/005 GUN

COST IN DOLLARS:	**Unknown**
CALIBER:	**35 mm**
DIMENSIONS WHEN IN OPERATION:	
Length	7.87 m
Height	2.90 m
Width	2.26 m
Length of barrel	3.15 m
WEIGHTS:	
Total in combat order	7,500 kg

FEATURES:	
Maximum range	2.5 miles
Maximum speed	50 mph
Maximum slope climbable	30 %
Maximum depth fordable	65 cm
Number of cartridges	65
Firing rate	1,200 shots per minute
CREW:	**3 men**

TABLE OF CONTENTS